THE LIVES OF

SEAWEEDS

THE LIVES OF | SEAWEEDS

A NATURAL HISTORY OF OUR PLANET'S SEAWEEDS & OTHER ALGAE

Julie A. Phillips

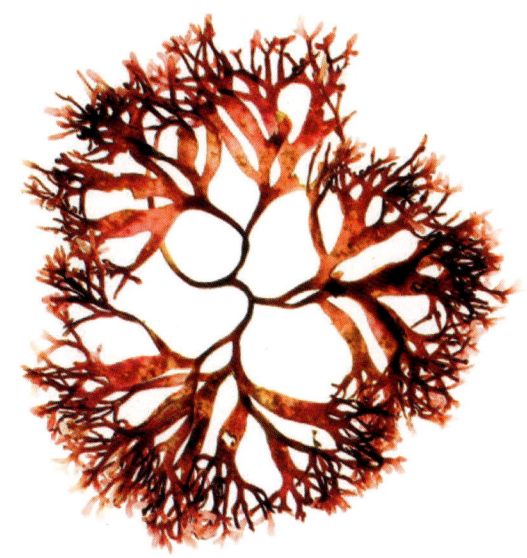

PRINCETON UNIVERSITY PRESS
PRINCETON AND OXFORD

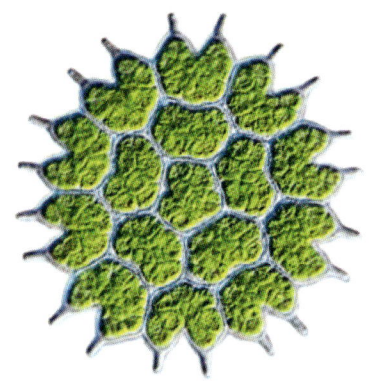

Published by Princeton University Press
41 William Street, Princeton, New Jersey 08540
99 Banbury Road, Oxford OX2 6JX
press.princeton.edu

Library of Congress Control Number 2023938099
ISBN 978-0-691-22855-6
Ebook ISBN 978-0-691-23017-7

Typeset in Bembo & Futura

Printed and bound in China
10 9 8 7 6 5 4 3 2 1

British Library Cataloging-in-Publication Data is available

This book was conceived, designed, and produced by
UniPress Books Limited
Publisher: Nigel Browning
Commissioning editor: Kate Shanahan
Project managers: Kate Duffy and Kathleen Steeden
Designer & art directon: Wayne Blades
Illustrator: John Woodcock
Maps: Les Hunt

Cover images: (Front cover) Brent Durand / Getty Images;
(back cover and spine) Steve Speller / Alamy

CONTENTS

The algal world

Loved by the Japanese, seaweeds have long been integrated into their cuisine, Shinto religion, and literature. Ladies of Victorian England made beautiful, pressed seaweed pictures and albums. Delicious sushi rolls with their nori seaweed wraps are a twenty-first-century global culinary obsession. The Red Sea and Sargasso Sea are named after algae. Seaweeds and other algae have a multitude of uses as food, in industry, and in medicine. Since time immemorial, algae have enriched the lives of humans, but these endlessly fascinating organisms also play a pivotal role globally in sustaining life on Earth.

Since prehistoric times, coastal people worldwide have foraged along their seashores harvesting seaweeds for food. Nowadays, the nutritious seaweeds are still eaten in many countries, more so in Japan, China, and Korea than in Norway, Britain, the United States, and Canada. Seaweeds have health benefits far beyond their macronutrient content, with some containing 10 to 100 times more minerals and vitamins per unit dry weight than terrestrial plants and animal-derived foods. Furthermore, dietary studies have reported significantly fewer obesity and diet-related diseases in Japan where seaweeds are regularly consumed, suggesting a strong link between seaweed consumption and good health.

Not only do the Japanese eat a larger quantity of seaweeds than other nationalities, they also ingeniously use them to enhance the flavors of other foods. Dashi stock, made from the fish, bonito, and the kelp, konbu, provides the sensational fifth (after sweet, sour, salt, and bitter) taste of umami, the "deliciousness" that underpins the famous Kyoto cuisine.

Several seaweed species are important in Shinto ceremonies and festivals, offered as food to the kami (gods). It is a Shinto belief that shaking, touching, and eating the brown seaweed *Sargassum* cleanses and purifies the human body. Interestingly, modern science has found that *Sargassum* and some other large brown seaweeds contain many bioactive chemical compounds that have potential health benefits owing to their antioxidant, anticancer, antibacterial, and anti-inflammatory properties.

The high regard the Japanese have for seaweeds is also apparent in their literature, in which seaweeds are used as metaphors to express feelings of love, compassion, truth, and sensuality. Even the famous *Tale of Genji*, written by the noblewoman Murasaki Shikibu in early eleventh-century Kyoto, possibly the first novel in world literature, has a romantic poem that mentions seaweed:

> *The world of the fisher folk*
> *Might I hear it from afar?*
> *The beach at Suma,*
> *Seaweed-salt droplets fell,*
> *For who, if not you . . .?*

→ Title pages in Victorian-era seaweed albums were often decorated with a poem and a frame of artistically arranged seaweeds. During pressing and drying, the mucilage in their cell walls glued the seaweeds onto the paper sheets.

FLOWERS OF THE SEA.

CALL us not weeds, we are Flowers of the Sea,
For lovely, and bright, and gay-tinted are we;
And quite independent of culture or showers;
Then call us not weeds, we are Ocean's Gay Flowers.

SEAWEEDS AND OTHER ALGAE

Visible to the unaided eye, seaweeds, also called marine macroalgae, are the familiar macroscopic algae that inhabit rocky seashores. Less well known are the pondweeds, the macroalgae of freshwater lakes, rivers, ponds, and streams. Seaweeds and pondweeds are lumped together by their habitat, but actually they belong to five different evolutionary lineages partially distinguished by the dominant pigment that colors the alga blue-green (phylum Cyanobacteria), red (phylum Rhodophyta), green (phylum Chlorophyta, phylum Charophyta), or brown (class Phaeophyceae).

The body of a macroalga (seaweed and pondweed) is called the thallus; it has a simple structure. In most macroalgal species, one part of the thallus resembles the rest, as is evident in the netlike thallus of the red seaweed *Claudea elegans*. Notable exceptions are some large brown seaweed species that have rootlike, stemlike, and leaflike structures composed of similar types of tissues. Macroalgae do not have true roots, stems, and leaves. These structures evolved in the land plants as adaptations to life on land and are composed of highly differentiated tissue types, which are involved in transporting water (xylem) and sugars (phloem) around the plant, providing structural support, and reducing water loss.

Seaweeds and pondweeds exhibit an amazing diversity in form, ranging from small, barely visible filaments and crusts on rocks to the largest of the giant leathery kelps 164 ft (50 m) long.

In addition to the macroalgae, there are far greater numbers of microscopic, unicellular algal species, often referred to as "microalgae." These species, which belong to several different evolutionary lines (lineages), are well represented in diverse habitats, most notably in the hidden invisible world of the phytoplankton, floating or swimming in the sea, and in inland waters. Even three phyla with many macroalgal species also contain microalgae. Numerous microalgal species are recorded for the green algae (phylum Chlorophyta and Charophyta) and the phylum Cyanobacteria, compared to the relatively small numbers for the Rhodophyta.

← This nineteenth-century hand-colored lithograph illustrates the exquisite netlike red seaweed *Claudea elegans*, which curls gracefully near its tips and whose reproductive structures (2–7) form within the net.

→ The firm hollow sac of the green bubble weed (*Dictyosphaeria cavernosa*) is formed of one layer of large cells clearly visible to the unaided eye.

WHAT IS AN ALGA?

The scientists who study algae are called phycologists. They are aware of the difficulties in defining the most megadiverse group of living organisms on Earth, well illustrated by the bizarre animallike unicellular alga *Michaelsarsia elegans*, whose tiny cell is covered by five different types of calcareous scales. For over two centuries, the algae had been defined as photosynthetic organisms that evolved in aquatic environments and, in the case of the macroalgae, never developed the roots, stems, leaves, and flowers that equipped the land plants for life in terrestrial environments. This definition, which is now known to have limited evolutionary significance, was reinforced by the two kingdom classification scheme that operated prior to the 1960s. In this scheme, the animals were assigned to the kingdom Animalia and all other living organisms (the land plants, algae, fungi, and bacteria) to the kingdom Plantae. It is now acknowledged that many algal groups are not closely related to the land plants or fungi.

By contrast, the land plants (the mosses, ferns, gymnosperms, and flowering plants) are genetically related to each other, a monophyletic group that originated from a common ancestor— a freshwater green alga. It is possible to trace the evolution of the land plants from a freshwater green algal ancestor through the mosses, ferns, and gymnosperms to the flowering plants, fulfilling the requirement that biological classification must be based on the evolutionary history of the organisms.

During the 1960s, a few biologists were fervently challenging the validity of the two-kingdom classification scheme. Fortuitously, two new techniques, electron microscopy (or ultrastructural) and DNA molecular studies, began generating exciting and new irrefutable evidence, ultimately proving two decades later that the algae had evolved from different ancestors and therefore are polyphyletic. From the 1960s to the 1990s, several new classification schemes were proposed, but most were subsequently found to have marked shortcomings until one, finalized in 1998 and widely accepted since, recognized six kingdoms of life for the planet.

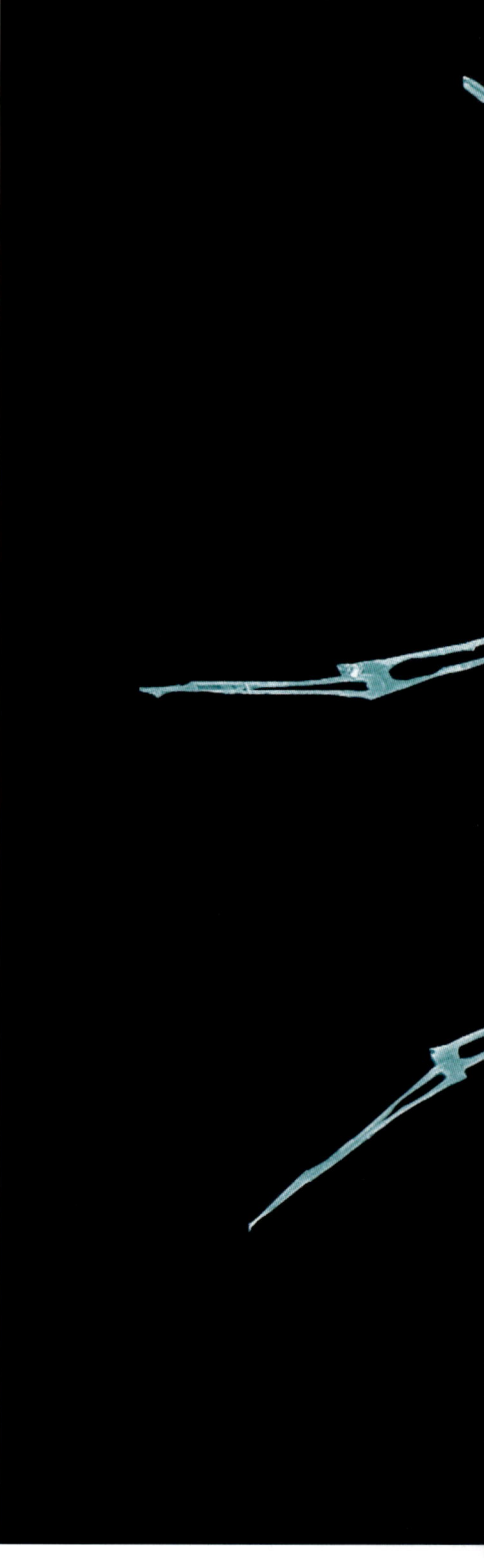

→   False-colored electron microscope image of the golden-brown alga *Michaelsarsia elegans*.

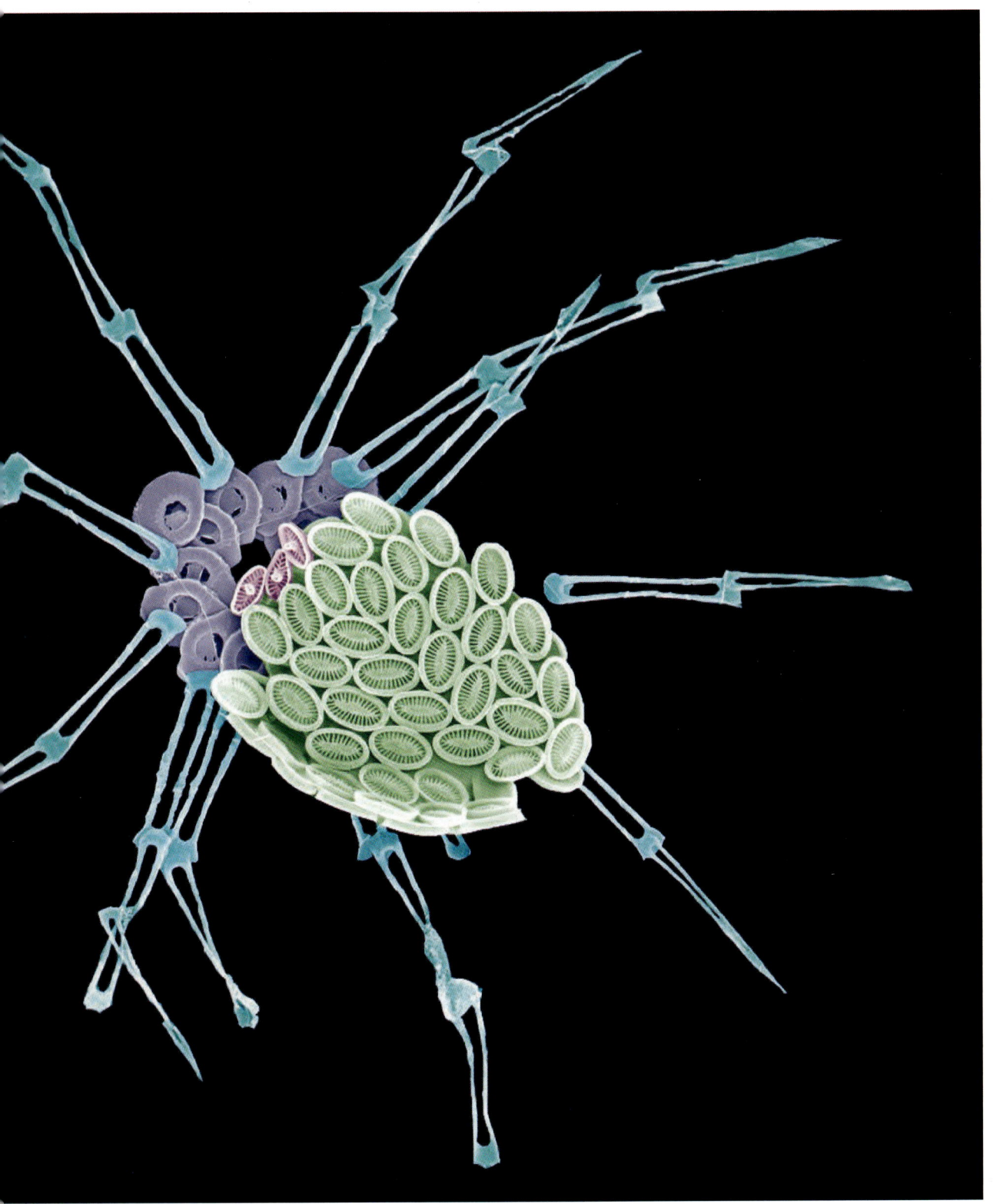

FOUR OF SIX KINGDOMS

The new classification scheme radically changed the higher (kingdom to class) levels of classification of the algae, assigning algal species to four of the six kingdoms. The blue-green algae are actually blue-green bacteria, the only algal phylum classified in the kingdom Bacteria. The red algae (Rhodophyta) and green algae (Chlorophyta and Charophyta) are plants and are assigned to the kingdom Plantae; the euglenoids, a group of unicellular green microalgae, to the kingdom Protozoa; and the brown algae (Phaeophyceae) and many microalgal groups to the exclusively algal kingdom Chromista. Brown seaweeds that include the spectacular *Tomaculopsis herbertiana* are the most familiar members of the kingdom Chromista. Only the kingdoms Animalia and Fungi lack algal species.

The classification of the algae is rigorous, based on several independent lines of evidence derived from morphological, biochemical (pigment composition, storage carbohydrate, and cell wall composition), ultrastructural (structure of the cell wall, motile cells), and, since the 1990s, DNA sequence data. Morphological and anatomical characters are observed with the light microscope at magnifications to 1,000

← The brown seaweed *Tomaculopsis herbertiana* is spectacularly iridescent. Each tuft of radiating hairlike filaments emits gorgeous hues of brilliant blue and green.

WHAT ARE PHOTOTROPHS, HETEROTROPHS, AND MIXOTROPHS?

Living organisms are often described by their mode of nutrition as phototrophs, heterotrophs, or mixotrophs. Long characterized as photosynthetic organisms, the majority of algal species are indeed phototrophs, defined as deriving their nutrition from photosynthesis: a process that produces the organic compound glucose by directly capturing the sun's energy. The growth of phototrophs is largely limited by temperature and the availability of light and the nutrients nitrogen and phosphorus.

However, not all algae are phototrophs. Some algal species are heterotrophs. They are incapable of photosynthesis and derive their energy from consuming organic compounds or other living organisms. Still other algal species are mixotrophs, capable of both phototrophic and heterotrophic modes of nutrition, the balance between these modes also varying. Some mixotrophs switch to absorbing organic matter when the light is too low for photosynthesis, while others need to both eat other organisms and photosynthesize to survive.

The algae are unique in having species classified in four of the six kingdoms of life. Algal species exhibit a great diversity in form, from the weird to the wonderfully beautiful. They live in almost every habitat on Earth, from the stable and favorable subtidal habitats in shallow seas to inhospitable acidic hot springs. Various algal species drive the global biogeochemical cycles that make the planet habitable. This book provides a snapshot of the exciting megadiverse world of the algae.

times and ultrastructural characters with the electron microscope at magnifications to 100,000 times.

The number of algal species currently existing on Earth remains unknown. An inventory lists 32,260 of an estimated 43,918 described algal species, a number that many phycologists believe is far from complete. They consider that the known algal species will increase by a factor of four to eight once poorly known geographical regions and habitats are surveyed and after cryptic species, defined as several morphologically similar species masquerading as one, are identified by DNA sequencing studies.

EVOLUTION

Billions of years of algal evolution

How phycologists unraveled the amazing evolutionary history of the organisms traditionally assigned to the algae is a remarkable story. The evolution of the algae—and indeed all plants—began around 3.5 billion years ago, when cyanobacteria (formerly called blue-green algae) were among the first living organisms on Earth.

THE ARCHEAN LANDSCAPE

Imagine being transported back 3.5 billion years in time into the landscape of the Archean eon. Having originated 1 billion years earlier, Earth was a harsh environment that was inhospitable to life as we know it today. Water was abundant and had formed the oceans. There was a significant amount of volcanic activity and the atmosphere was composed mostly of nitrogen and carbon dioxide, with little or no oxygen. At this time, bacteria that used substances other than oxygen for respiration were the only life on the planet capable of surviving in the low-oxygen environment.

In this primeval soup, one type of bacteria—the cyanobacteria—would have a profound effect on the evolution of life. Cyanobacteria are responsible both for releasing oxygen into the early atmosphere and for the evolution of all organisms on Earth that possess plastids. Plastids are membrane-bound structures (organelles) found in the cytoplasm of the majority of algal and plant cells and, importantly, are the site of photosynthesis. Uniquely positioned as the first oxygenic photosynthetic organisms on Earth, the cyanobacteria synthesized glucose and liberated oxygen into the atmosphere as the by-product. Around 2 billion years ago, cyanobacterial photosynthesis had expelled so much oxygen into the atmosphere that it could support the evolution of nonbacterial life. It might sound like the premise of a science fiction movie, but around this time a colorless cell engulfed and took a cyanobacterial cell into its cytoplasm, where it became a plastid. By acquiring a plastid, the colorless cell was transformed into the first plant cell on Earth that

subsequently gave rise to three algal lineages. This was not a one-off occurrence; it happened again, many times over, some as recently as 250 million years ago. On these occasions, colorless cells engulfed and enslaved either a red or a green algal cell, which became the plastids for six more algal lineages. It is amazing to think that plastids not only played a pivotal role in the evolution of life on Earth, but through photosynthesis have driven a range of important natural processes and helped shape the global economy.

↑   The first life-form on Earth, the moundlike stromatolites evolved in Earth's inhospitable aquatic environments in the Archean eon 3.5 billion years ago.

Symbiotic theory of algal evolution

Different types of plant plastids have long been recognized. The well-known chloroplasts of the land plants are different to the red plastids of the red algae, the brown plastids of the brown algae, and the colorless plastids of plant roots and plant parasites. However, the full extent of plastid diversity and the pivotal role that plastids played in algal evolution has only been realized in the last five decades.

SYMBIOSIS

As early as the 1880s and again in the early 1900s, some botanists proposed the radical theory that plastids had originated by symbiosis, a situation in which two different organisms live together in close association. In this symbiosis, a cyanobacterium (as a plastid) lives inside a colorless cell; both the cyanobacterium and the cell benefit from the association. The cyanobacterium receives protection, carbon dioxide, and nutrients (nitrogen and phosphorus) from the cell, which, in turn, receives the products of photosynthesis (glucose and oxygen) from the cyanobacterium.

Their symbiotic theory would take 100 years to prove. The early evolution of the algae had taken place inside cells and could only be investigated from the late 1950s after the invention of electron microscopy and DNA techniques. Over the following decades, these new techniques began accumulating evidence piece by piece that documented the differing structure and the genetic relationships of algal plastids.

Plastids have long fascinated botanists. Between 1846 and 1885 they reported the astounding observations that the plastids of various algal species proliferated by dividing into two (the same pattern of cell division—binary fission—that characterizes the bacteria), rather than dividing when their cell divided by mitosis. Botanists also suggested that plastids were passed from generation to generation through the female gamete. This information led the German ecologist A. F. W. Schimper to surmise in his 1883 paper that plastids and their host cells were somewhat symbiotic, and that green plants may have originated through the "unification of a colorless organism with one uniformly tinged with chlorophyll."

In 1905, Constantin Mereschkowsky, a Russian botanist from a small German university, proposed that plant cells acquired their plastids when a photosynthetic bacterium was engulfed by an animal cell, giving rise to the three main branches in the plant kingdom: the red, green, and brown algae. Mereschkowsky not only compared plastids to "little green slaves" that harnessed sunlight to provide their hosts with energy, but also identified other aspects of the symbiosis. He observed that the plastids were transmitted from generation to generation, continued to function in cells whose nucleus had been removed, lived in the cells of animals such as *Amoeba* and *Hydra*, and resembled the lower forms of cyanobacteria. These were astute observations, made at a time when laboratory equipment was primitive and more than 25 years before the vast differences between bacterial and nonbacterial cells were first proposed. However, while Mereschkowsky's symbiotic theory was initially accepted, after World War I it was dismissed as being based on wild speculation. Relegated to a fringe hypothesis, it was largely neglected for decades to come.

← Microscopic plastids, including the star-shaped chloroplasts in cells of the freshwater green alga *Zygnema*, have played a central role in algal evolution.

Cyanobacteria and stromatolites

The important role that cyanobacteria played in the evolution of early life on Earth has been preserved in the fossil record. The ancient cyanobacteria that inhabited shallow marine environments grew in mats, forming layered microbial communities called stromatolites. The 3.5-billion-year-old stromatolites found in the Archean rocks in Western Australia's Pilbara region—the oldest identifiable fossils on Earth—are crucial to our understanding of the origins of life on the planet.

LIVING FOSSILS

Although they were rare in the Archean eon, stromatolites became more abundant when shallow marine environments expanded around 2.5 billion years ago. About 850 million years ago, they became the dominant life form on Earth, peaking in abundance before experiencing a sharp decline some 570 million years ago but persisting until the present. Modern stromatolites are constructed by cyanobacteria as well as other more recently evolved algae. Today, these living fossils can be found in the United States, the Bahamas, the Persian Gulf, the Red Sea, and Australia. However, their distribution is often restricted to extreme environments that are similar to those of the Archean eon, such as the thermal springs of Yellowstone National Park, the hypersaline Great Salt Lake in the United States, and Hamelin Pool in

Australia's Shark Bay, where the toxic salinity is around twice that of seawater.

Cemented to the underlying rock platform, stromatolites are dark-green, stony, domed mounds that reach around 2 ft (60 cm) in height—some project above the water's surface, while others grow fully submerged. These organo-sedimentary structures are formed when the gluelike mucilage produced by the mat-forming cyanobacteria and the other algae living in the spongy upper surface (the organic

component) trap, bind, and deposit sediment. The cyanobacteria and other algae glide on the mucilage and migrate toward the light, always remaining on the sunlit upper surface of the stromatolite in order to photosynthesize, and to avoid burial by the deposited sediment. Most stromatolites grow extremely slowly. In Hamelin Pool, those with a growth rate of less than 0.04 inches (1 mm) per year have taken hundreds of years to achieve their maximum height of 2 ft (60 cm).

↑　The stromatolite in hypersaline Hamelin Pool, Shark Bay, Western Australia, is not an inanimate rock but a living structure with a soft and spongy upper surface.

↖　The ultimate living fossil, strange moundlike stromatolites have been made by the cyanobacterial inhabitants of aquatic environments for the last 3.5 billion years.

Prokaryotes and eukaryotes

The Cyanobacteria are similar to other algae in their biochemistry and physiology. All algal groups are defined in part by the possession of the green pigment, chlorophyll *a*, and the capacity for photosynthesis. Cyanobacteria are the only bacteria with these characteristics. Furthermore, the Cyanobacteria and the red algae share other biochemical and physiological characteristics including blue-green and red photosynthetic pigments.

STRUCTURAL DIFFERENCES

Despite these similarities between the Cyanobacteria and the red algae, profound differences in the cell structure exist between bacteria—including cyanobacteria—and all nonbacterial life (chromists, plants, animals, and fungi). The differences are far greater than those that exist between plant and animal cells: bacteria are prokaryotes, whereas all other living species are eukaryotes.

PROKARYOTIC CELLS

Prokaryotic cells have a simple level of organization. There are no membrane-bound structures inside the cell. There is no nucleus, only strands of DNA. The cyanobacteria do not have plastids, only many thylakoid membranes—the sites for photosynthesis—on which are situated spherical or disklike structures containing the photosynthetic pigments.

The cyanobacterial cell wall consists of outer and inner membranes that are separated by a carbohydrate and protein middle layer. There is no sexual reproduction. Instead, prokaryotic cells divide into two equal halves through the process of binary fission (a typically bacterial method of cell division).

EUKARYOTIC ALGAE

In contrast, the eukaryotic cell has a higher level of organization, which is evident by the membrane-bound structures within it: the nucleus is bounded by two nuclear membranes (the nuclear envelope), the photosynthetic thylakoid membranes are organized into a plastid bounded by plastid membranes (the plastid envelope) and cellular respiration occurs in membrane-bound mitochondria. Unlike the nuclei and mitochondria, which are remarkably uniform structures, the structure of the plastid varies among the algal phyla. The number of membranes in the plastid envelope varies from two to four and the number of thylakoid membranes stacked in the plastid matrix generally varies from one to six. These characters are important in understanding the evolution of the algae. Cells divide by mitosis and produce gametes for sexual reproduction.

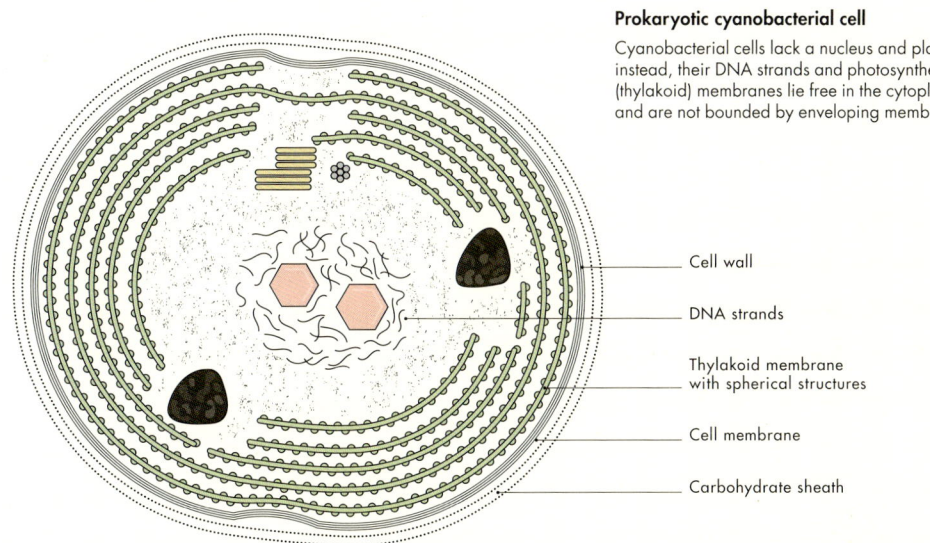

Prokaryotic cyanobacterial cell

Cyanobacterial cells lack a nucleus and plastids;
instead, their DNA strands and photosynthetic
(thylakoid) membranes lie free in the cytoplasm
and are not bounded by enveloping membranes.

Cell wall

DNA strands

Thylakoid membrane
with spherical structures

Cell membrane

Carbohydrate sheath

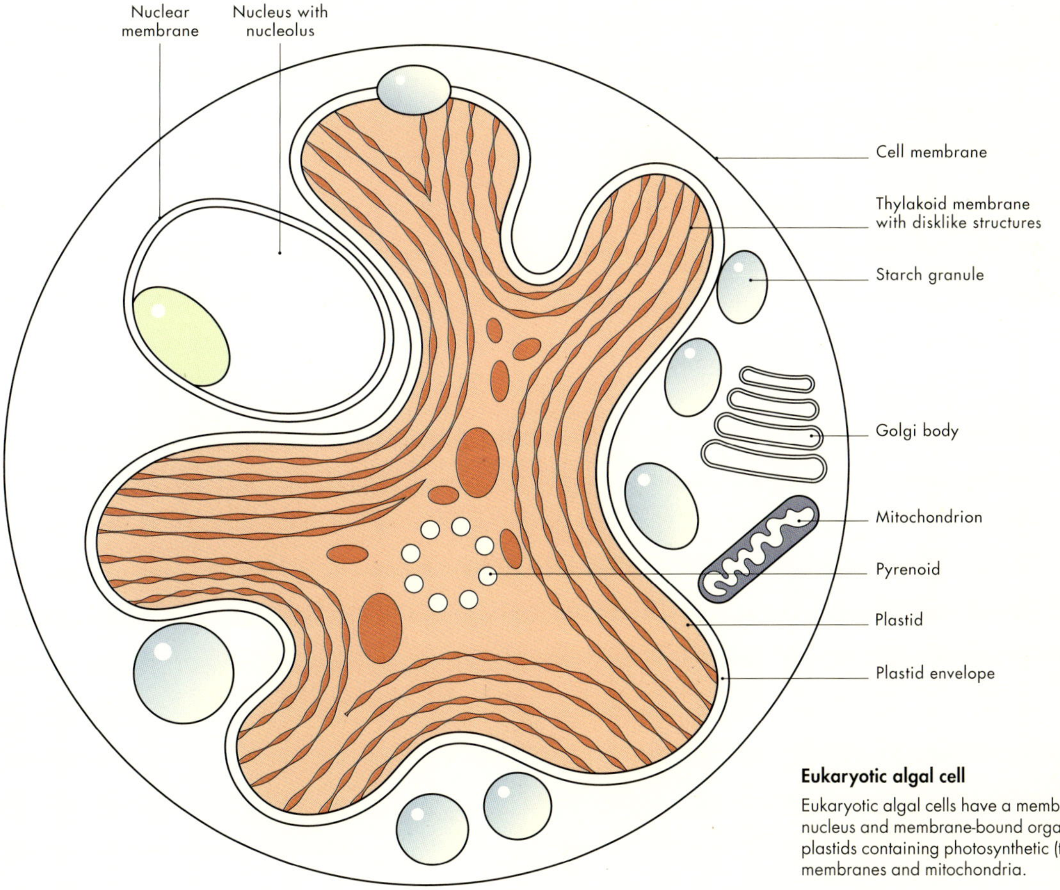

Nuclear
membrane

Nucleus with
nucleolus

Cell membrane

Thylakoid membrane
with disklike structures

Starch granule

Golgi body

Mitochondrion

Pyrenoid

Plastid

Plastid envelope

Eukaryotic algal cell

Eukaryotic algal cells have a membrane-bound
nucleus and membrane-bound organelles, including
plastids containing photosynthetic (thylakoid)
membranes and mitochondria.

The symbiotic theory resurrected

The resurrection of the symbiotic theory was largely due to the efforts of the American biologist Lynn Margulis, of Boston University. Based on cytological, biochemical, and paleontological evidence, Margulis revitalized the proposal that plastids and mitochondria arose through symbiosis. Attracting much skepticism, her radical hypothesis was eventually published in 1967 (under the name Lynn Sagan), having been rejected by 15 scientific journals.

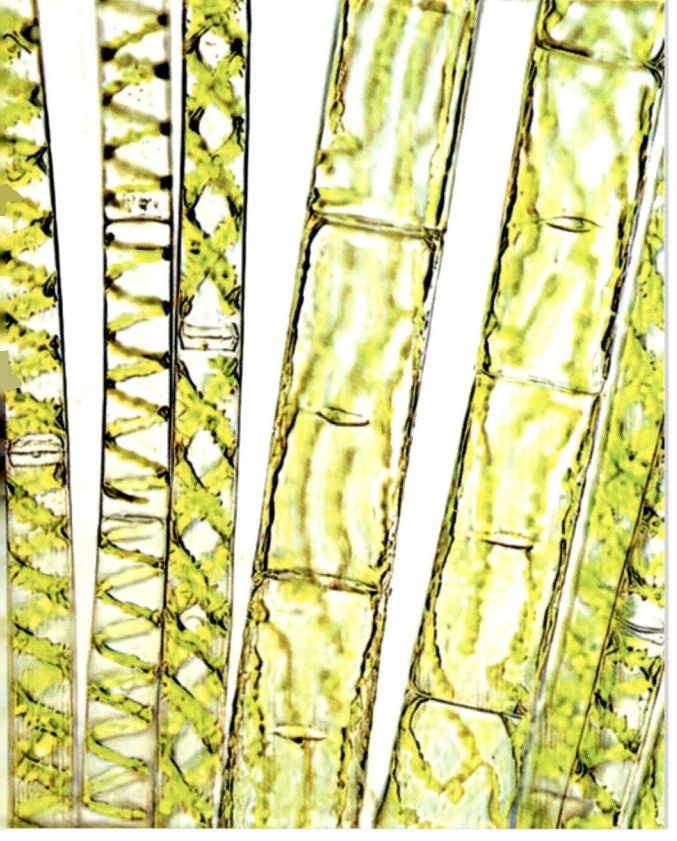

THE EVIDENCE

The symbiotic theory had remained out of favor until 1959, when DNA was discovered in the plastid of the green alga *Spirogyra*. Until then, DNA—the genetic material—was thought to only occur in the nucleus. This amazing discovery reignited interest in the symbiotic theory, although some biologists continued to challenge its validity.

More startling revelations followed. *Spirogyra* was only one of many algal and land plant species whose plastids were found to contain DNA. Then, in 1974, great excitement resonated throughout the botanical world when electron microscopy studies revealed the presence of a reduced nucleus in the plastid of a single-celled alga (a cryptophyte).

↗ In the 1960s, biologist Lynn Margulis (1938–2011) began championing, against much early opposition, the new radical theory of algal evolution.

← The 1959 discovery of DNA in the large, ribbonlike chloroplasts of the green alga *Spirogyra* reignited interest in the central role that plastids have played in the evolution of the algae.

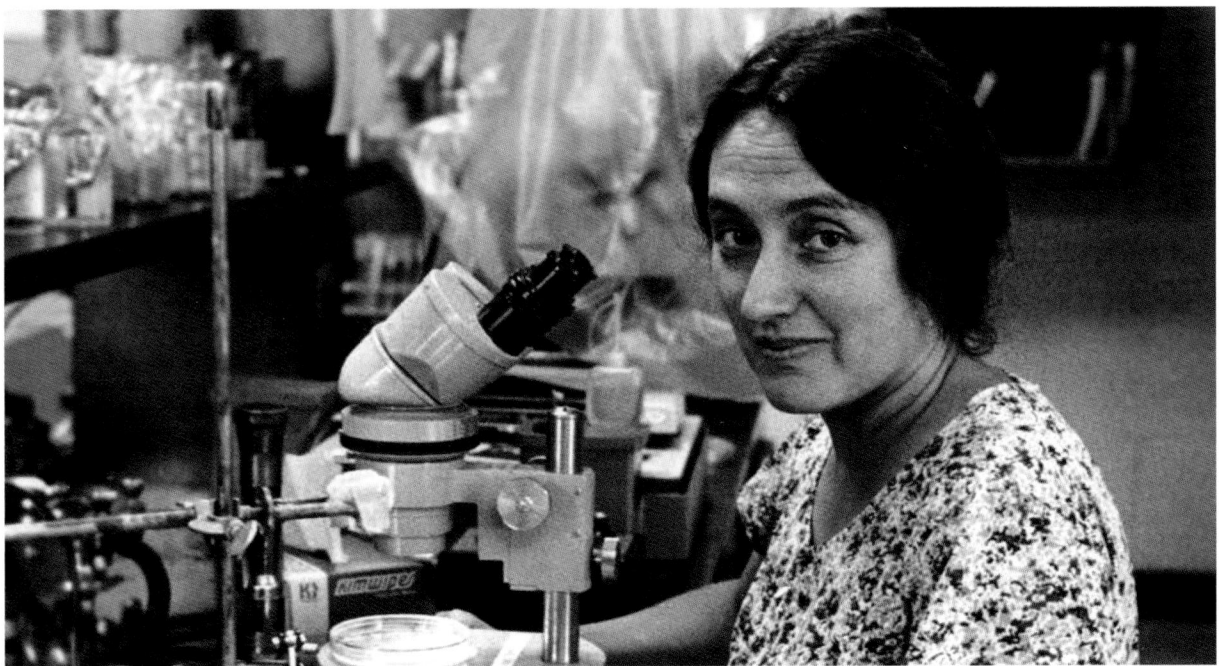

The evidence supporting the symbiotic theory continued to accumulate particularly when DNA studies established that the plastid genes in the primitive red algal genus *Porphyra* were most closely related to the genes of the cyanobacteria. The mounting evidence suggested that the plastids of the algae and land plants had originated from a cyanobacterial cell.

THE SYMBIONTS

A huge research effort over the last five decades has established how algal cells acquired their plastids by endosymbiosis, the process by which one symbiont lives inside the other symbiont ("endo" means "inner"). The first plastids of algal cells were acquired when a colorless cell ingested a cyanobacterium that continued to live as an endosymbiont inside the host cell.

One cell ingesting another cell is a common method of feeding employed by unicellular organisms. In this process—known as phagocytosis—the cell membrane forms a pocketlike depression in which the food item is trapped. The membrane surrounding the food pinches off from the rest of the cell membrane, amazingly without disrupting the continuity of the cell surface, to form a membrane-bound food vacuole in the cell. Usually, the food item in the vacuole is digested. However, the endosymbionts that formed the first plastids escaped this fate.

MITOCHONDRIA

Mitochondria are the powerhouses of cells, the sites of metabolic processes including respiration. Mitochondria also originated by endosymbiosis. They were formed when a colorless cell engulfed an oxygen-consuming bacterium. Unlike the many different types of plastids, which were acquired on several occasions, the mitochondrion was acquired only once, given their relative uniform structure in the cells of animals, fungi, algae, and the land plants. This event pre-dated the acquisition of plastids by cells that subsequently gave rise to the algae and land plants.

The evolutionary saga begins

Around 1.5 billion years ago, a colorless cell was feeding when it ingested and enclosed a cyanobacterial cell in a membrane-bound vacuole in its cytoplasm. The cyanobacterial cell became a plastid that transformed the once colorless cell into an algal cell. This event—primary endosymbiosis— was the beginning of an evolutionary trajectory that would result in at least nine different algal lineages.

PRIMARY ENDOSYMBIOSIS

After its capture, the cyanobacterial cell became an endosymbiont that lived in its host's cell. Two cyanobacterial cell membranes envelope the endosymbiont, following the loss of the membrane produced by the colorless cell, to surround the ingested cyanobacterium. The endosymbiont was reduced significantly in size and a large number of its genes were transferred to the host's nucleus. These processes transformed the ingested cyanobacterium into a functional plastid, largely composed of photosynthetic thylakoid membranes. Remarkably, the plastid has been transferred from generation to generation in the host cell.

In primary endosymbiosis, the cell with its newly acquired plastid is the common ancestor to three algal lineages: blue-gray algae, red algae, and green algae. In these three lineages (and the land plants, which are descendants of the green algae), the two cyanobacterial membranes that bound the plastids is an important characteristic that places these organisms in the kingdom Plantae (see pages 32 and 278–279).

THE FIRST THREE ALGAL LINEAGES

The phylum Glaucophyta (the blue-gray algae) is a group of approximately 15 species of freshwater unicellular flagellates and colonies that form the missing link in the transition between the primitive cyanobacterium and the plastid of the red and green algae. The plastid of the blue-gray algae is bounded by an intact cyanobacterial cell wall with the two membranes separated by the middle protein and carbohydrate layer. Like cyanobacteria and red algae, glaucophytes have blue-green and red phycobilin photosynthetic pigments located in spherical or discoid structures on the surfaces of the single thylakoid membranes.

The phylum Rhodophyta (the red algae) is a large group of predominantly marine macroalgae (seaweeds). Red algal plastids have lost the middle protein and carbohydrate layer that is typical of the cyanobacterial cell wall, but retain the cyanobacterial characteristics of the blue-green and red phycobilin photosynthetic pigments, located in the spherical or disklike structures on the surfaces of single thylakoid membranes.

The phylum Chlorophyta (the green algae) is a large group of unicellular, colonial, or macroscopic plants that live predominantly in freshwater and marine habitats—the green seaweeds are the best-known species. Green algal plastids have lost many of the cyanobacterial characters that are present in the blue-gray and red algae and are characterized by stacks of two to many thylakoid membranes as well as the presence of a new pigment, chlorophyll *b*.

Primary endosymbiosis

From the top left: a colorless cell ingested and retained a cyanobacterial cell, which became the plastid (symbiont) of the first algal cell (top right). The first algal cell was the ancestor of the first three algal lineages: blue-gray algae, red algae, and green algae.

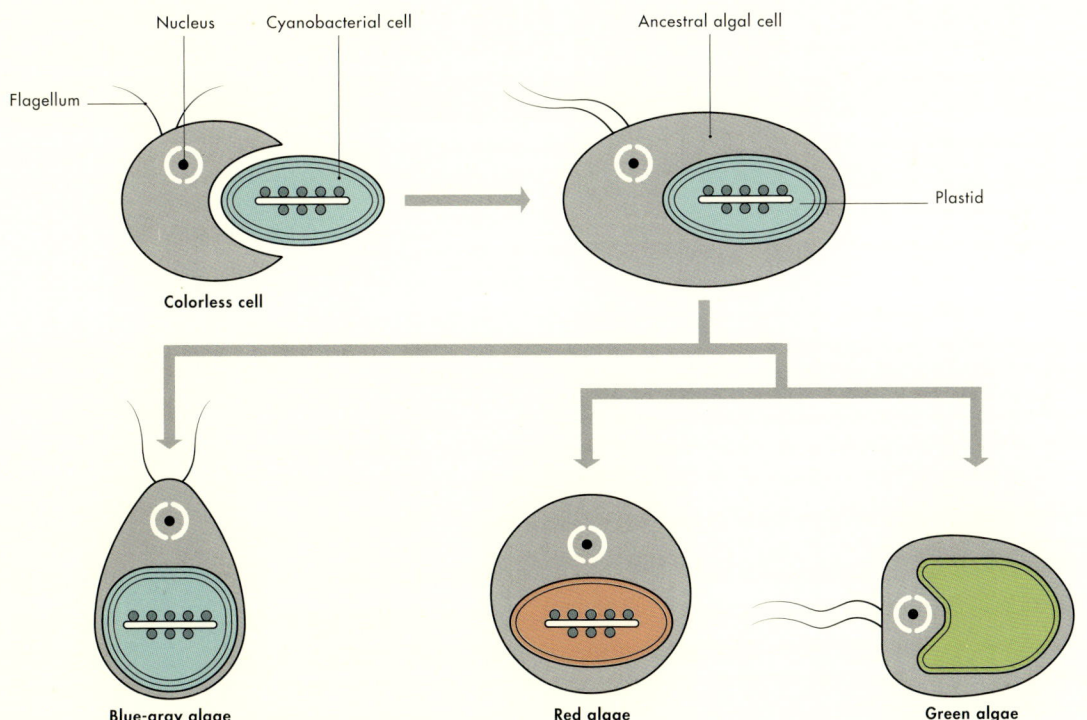

Flagellum

Nucleus   Cyanobacterial cell

Ancestral algal cell

Plastid

Colorless cell

Blue-gray algae

Red algae

Green algae

More algal lineages

More than a billion years after the events that resulted in primary endosymbiosis, it happened again, many times, some as recently as 250 million years ago. In this second round of endosymbiosis, a colorless cell ingested and retained a red or a green algal cell that had previously acquired their plastids from a cyanobacterial cell. The ingested red or green algal cell became the plastids of the new algal lineages.

FOUR PLASTID MEMBRANES

Six major algal lineages are recognized to have arisen through secondary endosymbiosis, two of which—the euglenoids (phylum Euglenozoa) and the green spider algae (phylum Chlorarachniophyta)—are derived from a green algal cell, with the remaining four derived from a red algal cell (see pages 32 and 278–279).

Secondary plastids—those that arise during secondary endosymbiosis—are usually bounded by four membranes: the innermost three originated from the ingested cell and the fourth from the colorless cell. The innermost three membranes originated from the two membranes of the plastid envelope and the cell membrane of the ingested green or red alga. The fourth (outermost) membrane was produced by the colorless cell and bounded the vacuole that took the red or green alga into the colorless cell. However, there are always exceptions in biology: the plastids of the euglenoids and dinoflagellates have only three membranes, having lost the cell membrane of the ingested green or red alga.

CHLOROPHYLL *B* ALGAE

Since the mid 1800s, photosynthetic pigments have played a central role in classifying the algae. There are several different kinds of the green pigment chlorophyll, which differ from each other only in their molecular structure. Chlorophyll *b* is found only in the green algae and the lineages that evolved from the green algae: the euglenoids, green spider algae, and the land plants. The euglenoids (phylum Euglenozoa) are a group of unicellular green flagellates that are common in, and largely restricted to, freshwater habitats. There are 2,000 known species, around half of which possess a plastid. The other half are colorless cells that are incapable of photosynthesis and feed either by absorbing soluble organic compounds or by preying on other unicellular organisms. Euglenoid plastids contain chlorophyll *a* and *b*, are bounded by three membranes, and have stacks of three closely appressed thylakoids. Now classified in the phylum Euglenozoa, the parasites that cause African sleeping sickness in humans (trypanosomes) are closely related to euglenoids.

The green spider algae are a small group of approximately 12 species of marine flagellates that include some species that are also capable of amoeboid movement. They are named for their green-colored plastids and the spiderlike shapes of their amoeboid cells. Their plastids contain both chlorophyll *a* and *b*, and are bounded by four membranes. A small and obscure group, the green spider algae gained enormous evolutionary importance with the discovery of a reduced nucleus in their plastids. A reduced nucleus had been discovered previously in another algal phylum. Only these two algal phyla are known to possess this structure in their plastid, which, in the case of the green spider algae, contains the genes from their green algal ancestor. This is another missing link that provides amazing evidence in support of endosymbiosis.

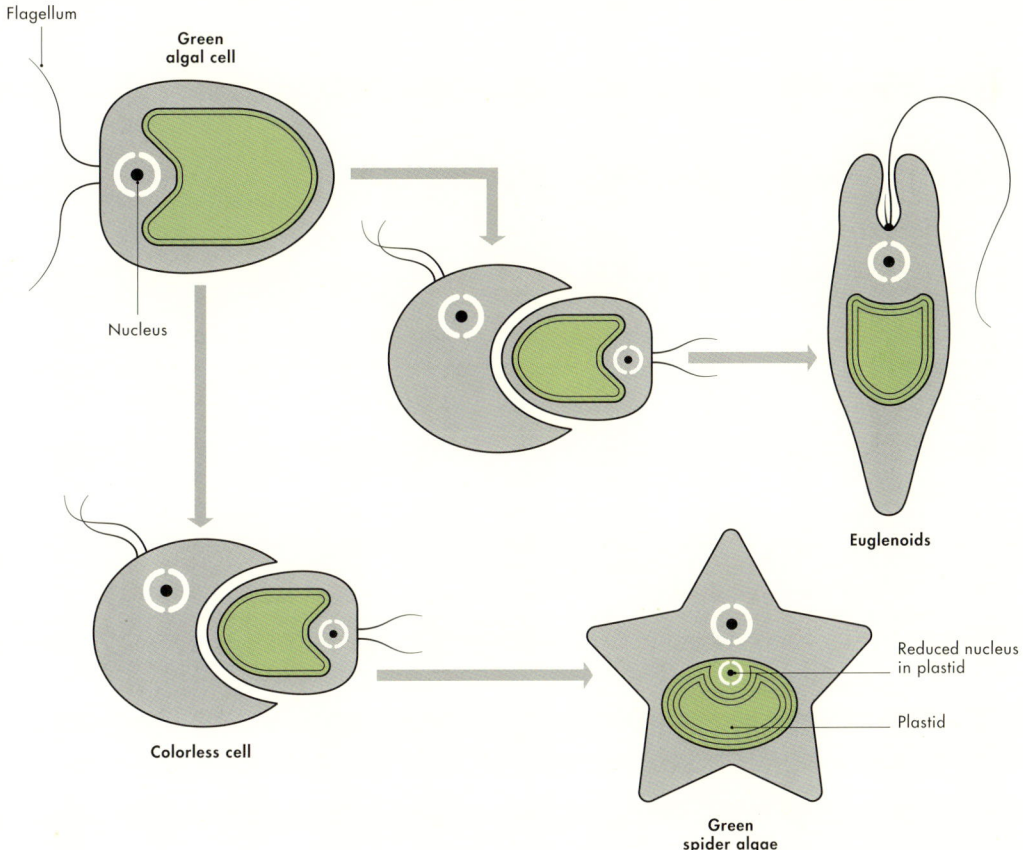

Flagellum

Green algal cell

Nucleus

Euglenoids

Colorless cell

Reduced nucleus in plastid

Plastid

Green spider algae

Secondary endosymbiosis: Green algal lineages

From the top left: Green algal cells formed by primary endosymbiosis are ingested and retained as plastids by two different types of colorless cells. One colorless cell gave rise to the euglenoids and the other gave rise to the green spider algae.

Chlorophyll *c* algae

As well as chlorophyll *a*, species in four of the six major algal lineages recognized as arising through secondary endosymbiosis also possess chlorophyll *c*, an important character that classifies these four lineages in the exclusively algal kingdom Chromista. Described in 1989, this species-rich kingdom has four lineages (Cryptophyta, Haptophyta, Heterokonta, and a composite dinoflagellate lineage) that are derived from a red algal cell.

KINGDOM CHROMISTA

The Cryptophyta is a small group of approximately 200 described species of unicellular flagellates that inhabit marine and freshwater environments. The cryptophyte plastid contains chlorophyll *a* and *c*, the blue-green and red phycobilin pigments, stacks of two thylakoid membranes, and is bounded by four membranes. Reported in 1974, the discovery of the first reduced nucleus in the cryptophyte plastid took phycologists by surprise and resonated throughout the botanical world as an important missing link in algal evolution. Red algal ancestry for the cryptophytes is clearly demonstrated by the phycobilin pigments in the plastid, and a reduced nucleus containing red algal nuclear genes. The cryptophytes and green spider algae are the only algal phyla known to have a reduced nucleus in their plastids.

With approximately 510 species of marine flagellates, the Haptophyta are ecologically important photosynthetic organisms. Some haptophyte species form vast algal blooms in the world's oceans that are visible from outer space. Their plastids contain chlorophyll *a* and *c*, the brown pigment fucoxanthin, and have stacks of three thylakoid membranes.

The heterokonts form an extremely diverse group of some 18 classes and 12,500 photosynthetic and nonphotosynthetic species. These range in form from unicellular and colonial species to the brown seaweeds,

including the giant kelps that can grow to 100 ft (30 m) or more in length. The plastids of the heterokont and haptophyte species are similar.

Comprising approximately 2,500 marine and freshwater species, the dinoflagellates are a remarkably varied and complex group of unicellular flagellates. Half of the dinoflagellate species lack plastids and are heterotrophs, while the major plastid type contains the uniquely dinoflagellate red pigment, peridinin, and has three enveloping membranes, plus stacks of two or three thylakoids.

OTHER ENDOSYMBIOSES

Among photosynthetic organisms, the dinoflagellates exhibit the greatest diversity in their plastids. Their amazing ability to lose, replace, or gain new plastids is most probably related to their lifestyle as voracious predators on other unicellular organisms—a lifestyle that is uncommon in the algal world.

Four dinoflagellate genera do not have the typical three-membrane, peridinin-containing dinoflagellate plastid. Instead, they have plastids harvested from four very different lineages: a green alga (*Lepidodinium*), a haptophyte (*Karenia*), a diatom (*Kryptoperidinium*), and a cryptophyte (*Dinophysis*).

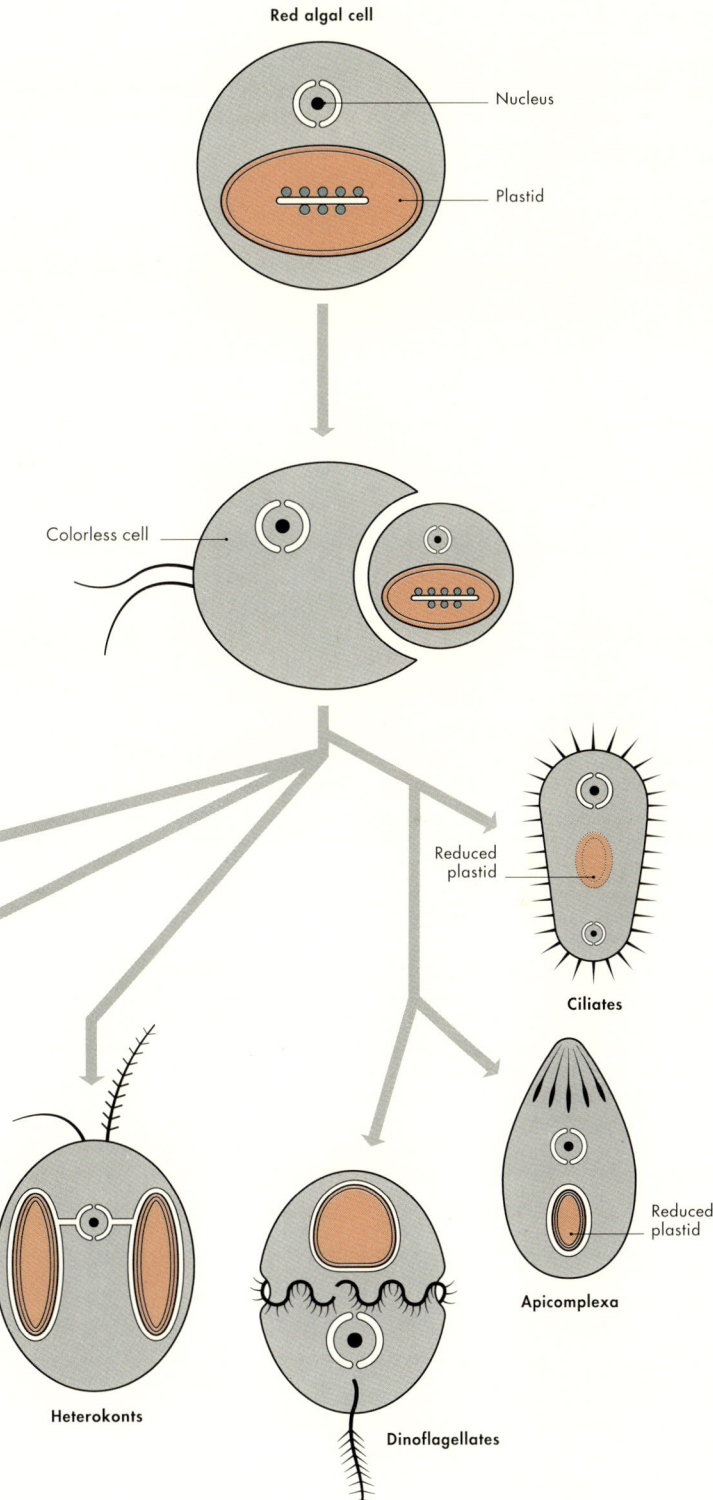

Red algal cell

Nucleus

Plastid

APICOMPLEXA

The Apicomplexa is a large group of intracellular parasites, including the malarial parasite, *Plasmodium*, and *Toxoplasma gondii*, which causes toxoplasmosis. The algal ancestry of Apicomplexa was recognized just 20 years ago, with the surprising discovery of the apicoplast (a relic plastid bounded by four membranes). The apicoplast genome resembles a reduced red algal genome that has lost all of the genes for photosynthesis.

Colorless cell

Reduced nucleus in plastid

Hairy flagellum

Plastid

Reduced plastid

Ciliates

Cryptophytes

Haptophytes

Reduced plastid

Apicomplexa

Secondary endosymbiosis: Red algal lineages

From the top: Red algal cells were ingested (below) and retained as plastids by four different types of colorless cells that gave rise to four algal lineages—the cryptophytes, haptophytes, the heterokonts, and a combined dinoflagellate, Apicomplexa, and ciliate lineage.

Heterokonts

Dinoflagellates

PHOTOSYNTHETIC PIGMENTS USED TO DEFINE SELECTED ALGAL PHYLA AND GROUPS

→ Cylindrical plants of the tropical green seaweed *Neomeris annulata* have heavily calcified (white) lower portions that dramatically increase this species' chances of preservation as fossils.

KINGDOM	PHYLUM/GROUP (COMMON NAME)	MAJOR PHOTOSYNTHETIC PIGMENTS	THALLUS COLOR
Bacteria	Cyanobacteria	Chlorophyll *a*, phycobilins	Blue-green or red
Plantae	Glaucophyta (blue-gray algae)	Chlorophyll *a*, phycobilins	Blue-green
Plantae	Rhodophyta (red algae, rhodophytes)	Chlorophyll *a*, phycobilins	Predominantly red, some blue-green
Plantae	Chlorophyta (green algae, chlorophytes)	Chlorophyll *a*, *b*	Green
Plantae	Charophyta (green algae, charophytes)	Chlorophyll *a*, *b*	Green
Protozoa	Euglenozoa (euglenoids)	Chlorophyll *a*, *b*	Green or colorless
Protozoa	Chlorarachniophyta (green spider algae; green amoebae)	Chlorophyll *a*, *b*	Green
Chromista	Crytophyta (cryptophytes)	Chlorophyll *a*, *c*, phycobilins	Red, blue-green, or colorless
Chromista	Haptophyta (haptophytes, coccolithophorids)	Chlorophyll *a*, *c*, fucoxanthin	Brown
Chromista	Dinophyceae (dinoflagellates)	Chlorophyll *a*, *c*, peridinin	Red or colorless
Chromista	Heterokonts (large group of 18 classes)	Chlorophyll *a*, *c*, some with fucoxanthin	Brown, golden-brown, or orange-brown

The first eukaryotic fossil

The oldest taxonomically resolvable multicellular eukaryotic fossil on Earth was found in 1.047-billion-year-old rocks on Somerset Island in Arctic Canada, and was identified as the red alga *Bangiomorpha pubescens*. Many ancient fossils that lack a distinctive morphology cannot be identified. Key to the identification of *Bangiomorpha* was its distinctive morphology, which is identical to only the extant red algal genus *Bangia*.

THE FIRST SEX

The red alga *Bangiomorpha pubescens* has the distinction of a triplet of firsts: not only is it the oldest eukaryotic fossil, but it is also the first occurrence of both complex multicellularity, and a sexually reproducing species in the fossil record. Bacteria living at the same time as *Bangiomorpha* did not reproduce sexually.

Eukaryote multicellularity, which introduced a macroscopic plant body, complex structure, and consequently an increasingly complex ecology, represents a critical threshold in the evolution of life on Earth. Although not the oldest multicellular eukaryotic fossil on record, *Bangiomorpha* holds the special distinction of true cellular differentiation represented by the possession of differentiated whole plants, with a holdfast that attaches the thallus to a rock, the capacity for the multiple cycles of cell division, and the ability to produce gametes.

Fortuitously, the distinctive morphology of *Bangia* distinguishes it from all other extant algal species. The plants are unbranched filaments measuring up to 4 in (10 cm) in length and up to 200 microns in diameter, with cells embedded in a firm, jellylike matrix. *Bangia* and *Bangiomorpha* are characterized by intercalary cell divisions that occur in the middle of the filament, not at its apex. In the young filaments with one row of cells, cell divisions elongate the filament and produce distinctive pairs of cells. Further cell divisions in the

young filaments produce filaments comprising many rows of cells, leading to many wedge-shaped cells positioned radially from the center of the filament to the filament surface. It is the shape and number of the cells in the young and older filaments that characterize and identify *Bangia* and *Bangiomorpha*.

Studying the fossil revealed two different types of reproductive cells resembling male and female gametes in separate wedge-shaped cells. These are similar to cells found in *Bangia*, indicating that *Bangiomorpha pubescens* not only reproduced sexually, but that sex contributed to the success of the eukaryotes in developing genetic recombination and multicellularity.

→ Filaments of the first eukaryotic fossil, *Bangiomorpha pubescens*, have distinctive paired vegetative cells (as seen in surface view), unique to *Bangiomorpha* (A) and its living relative *Bangia* (D), and distinctive wedge-shaped cells, as seen in cross section (B, C).

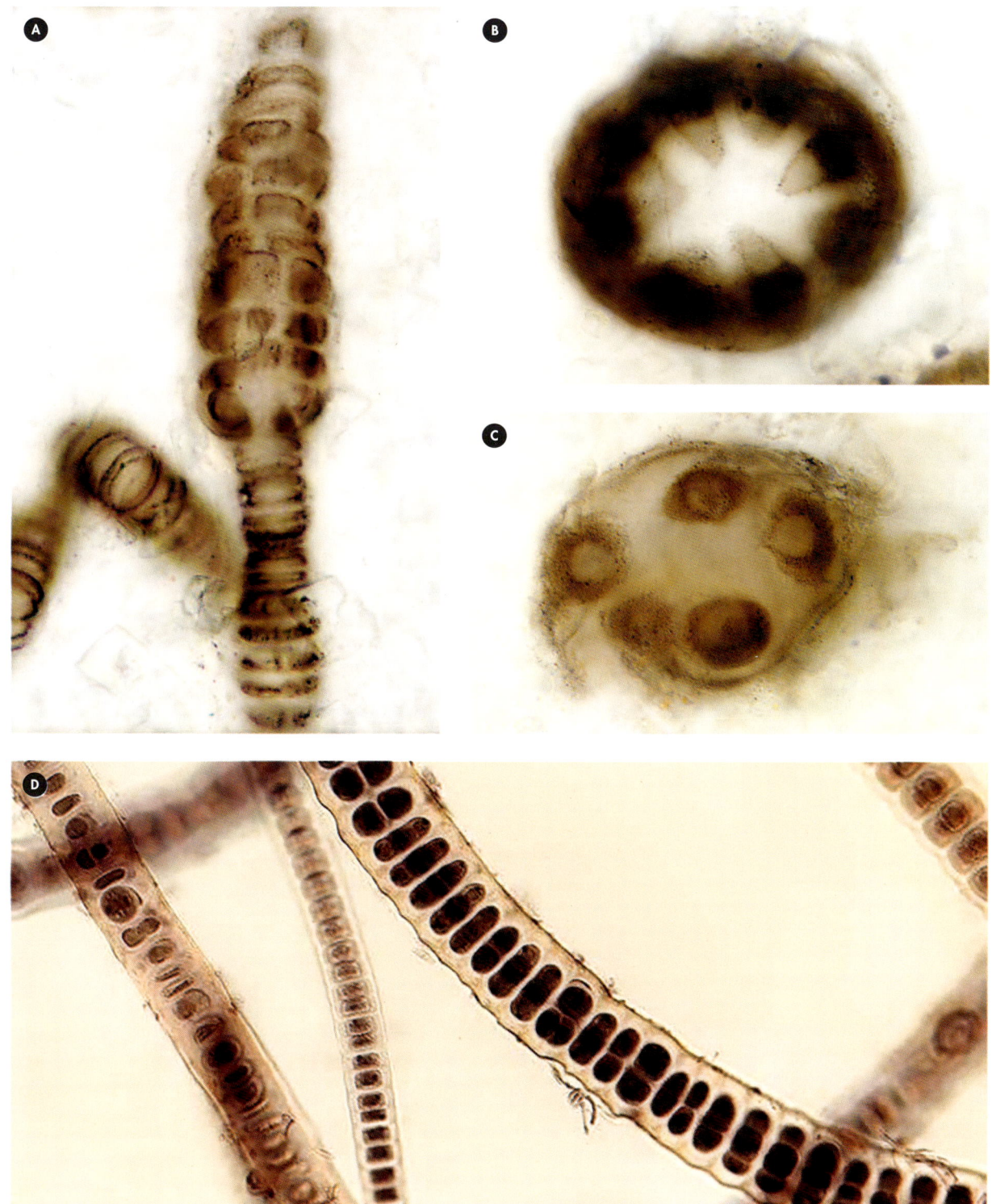

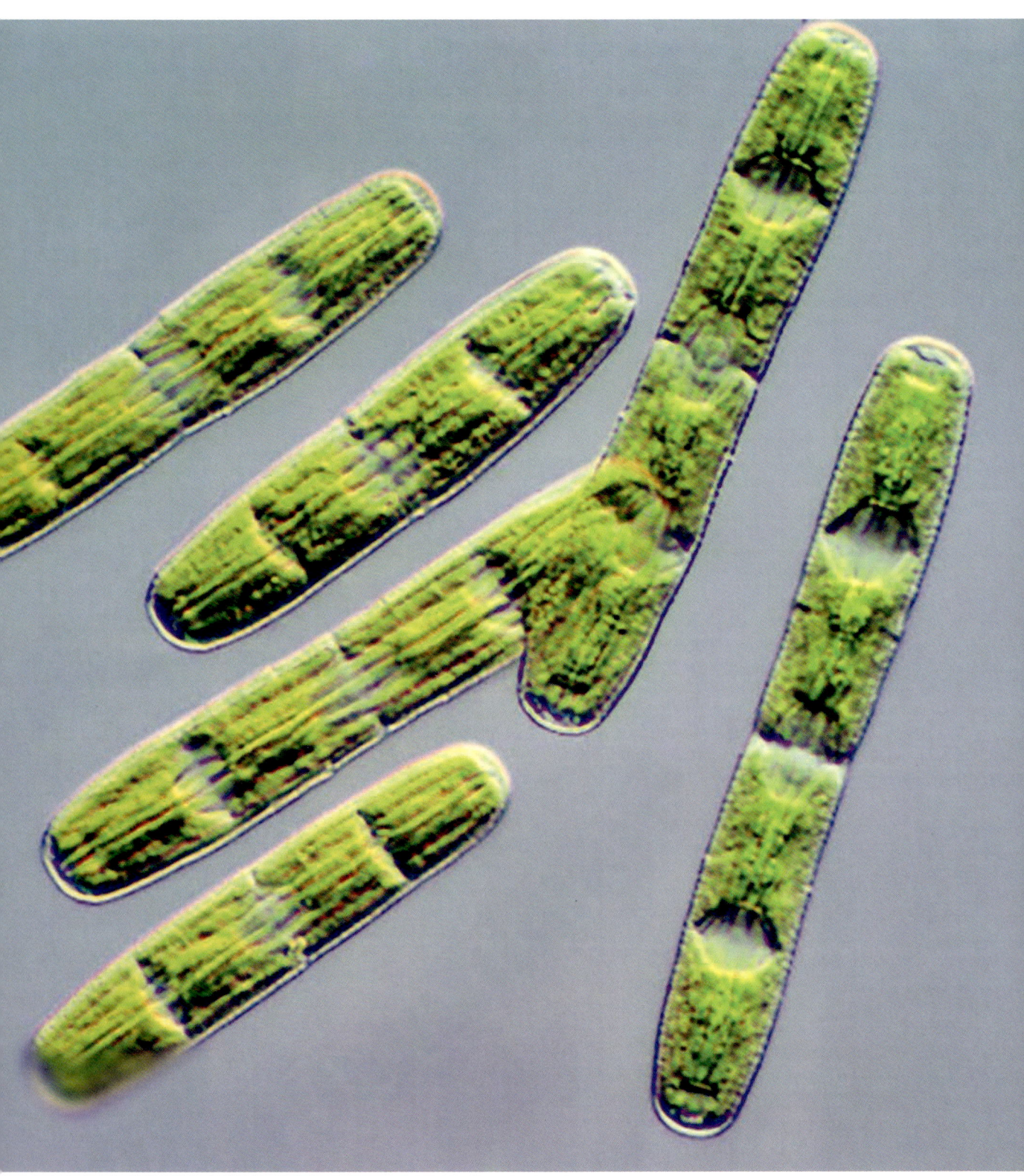

Ancestor of the land plants

It is hard to imagine Earth's bare terrestrial landscape before green plants emerged from freshwater environments to colonize the land in the mid Paleozoic era, around 430 to 470 million years ago. This pivotal event, and the subsequent radiation of the land plants into most terrestrial environments, had profound effects on the Earth's biogeochemistry, atmosphere, and biota.

THE CHAROPHYTES

The identity of the ancestors of the land plants remains elusive. There are no fossils of these plants and research on living green algal species has failed to provide insights into this evolutionary event.

However, the identity of the closest living relatives of the land plants is not disputed. It is not the green seaweeds and the many other species classified in the phylum Chlorophyta but another phylum in the highly diverse green algae, the largely freshwater Charophyta. In fact, the advanced charophytes—*Chara* and *Nitella*, commonly known as stoneworts—are more similar to the primitive land plants—the mosses and their close relatives—than to species of the Chlorophyta. The stoneworts and the primitive land plants have thalli that are morphologically similar, multicellular reproductive

← Inhabitants of peaty bogs and freshwater pools, desmid species, particularly *Penium margaritaceum*, are currently being investigated in the century-long search for the elusive ancestor of the land plants.

organs, structurally similar male gametes, the same pattern of cytoplasmic division during mitosis and several other shared characters. These traits are not shared with the chlorophyte algae.

Intriguingly, the stoneworts represent an endpoint in the evolution of the charophytes and the search for the common ancestor of the charophytes and the primitive land plants continues. The Charophyta comprises six lineages, but which one of these is most closely related to the ancestors of the land plants remains controversial. To date, three of these lineages have been sequentially considered (and discounted), with molecular studies currently proposing that the unicellular desmids are most closely related to the ancestors of the land plants. This theory is based on the premise that the desmids might occupy a habitat similar to that experienced by the ancestors of the early land plants. Their small size, rapid cell division, simple mode of sexual reproduction, and their ability to secrete copious amounts of water-retaining mucilage are favorable adaptations that enable the desmids to live at the interface between aquatic and terrestrial environments.

Algal fossil record

Algal species that possess calcified or silica structures are good candidates for preservation in the fossil record, while many other algal species composed of only soft tissues are rarely preserved. Despite this, the incomplete algal fossil record still provides insights into the origin and relationships of algal phyla, as well as supporting the molecular and ultrastructural data that underpins the theory of endosymbiosis.

KEY DATA POINTS

Dating back 3.5 billion years, the cyanobacterial fossil record is among the oldest for any group of organisms, confirming the antiquity of cyanobacteria. Having arisen 3.5 billion years ago, stromatolites were common on Earth around 2.3 to 2.4 billion years ago. This coincided with an increase in oxygenic photosynthesis (around 3.2

to 2.4 billion years ago) and, based on molecular data, pre-dates the establishment of algal cells by primary endosymbiosis (1.5 billion years ago). This correlates with the fossil record in 1.45-billion-year-old rocks that document the first appearance of decay-resistant, highly ornamented life history stages (acritarchs) that resemble modern prasinophytes, a primitive group of extant unicellular green algae. The occurrence of the multicellular (but still primitive) red alga *Bangiomorpha pubescens* in 1.047-billion-year-old rocks provides a key datum point for resolving eukaryote phylogeny: primary endosymbiosis, which gave rise to the red algae, must have occurred more than 1.047 billion years ago.

Molecular clocks variously date secondary endosymbiosis as occurring in the Haptophyta some 600 million years ago and in one heterokont group, the diatoms, no earlier than 240 million years ago. The fossil record dates the first appearance of some groups arising from secondary endosymbiosis to the Mesozoic era: 260 to 285 million years ago for the dinoflagellates, 220 million years ago for the Haptophyta, and 190 million years ago for the diatoms. The Haptophytes peaked in abundance during the Cretaceous, 65 to 93 million years ago, laying down extensive chalk layers across northern Europe, including the White Cliffs of Dover in England.

Calcareous red and green macroalgae have left a good fossil record. The uncalcified relatives of the modern coralline red algae arose 600 million years ago, with the calcified species appearing 100 million years later. The green algal order Dasycladales emerged during the early Cambrian, 520 million years ago, while the green algal genus *Halimeda* appeared during the Late Triassic, around 225 million years ago.

← The White Cliffs of Dover are formed from microscopic calcareous scales that covered the cells of unicellular haptophyte algae.

Algal classification in the twenty–first century

The realization in the 1970s that plastids originated by endosymbiosis was a major turning point in understanding algal evolution. Using extant species and the fossil record to unravel events that have occurred across 3.5 billion years is challenging. Nevertheless, evidence based on morphology, anatomy, biochemistry, fossils, ultrastructure, and molecular DNA studies has produced a six-kingdom classification for life on Earth.

RECONCILING CLASSIFICATION SCHEMES

Today, there is more than one biological classification scheme for all species on Earth, which largely reflects different approaches using a variety of techniques. Molecular biologists, for example, construct phylogenetic trees and deal with groups of organisms. One such phylogenetic tree divides all life into five super groups, including Plantae and Unikonts (which contains the fungi and animals, among other groups). However, reconciling phylogenetic trees with the traditional binomial classification system can sometimes prove difficult. This book follows the classification scheme devised by Professor T. Cavalier-Smith that is widely accepted by the majority of phycologists and which provides a comprehensive classification at all levels from kingdom to species.

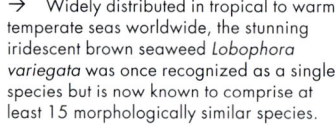

→ Widely distributed in tropical to warm temperate seas worldwide, the stunning iridescent brown seaweed *Lobophora variegata* was once recognized as a single species but is now known to comprise at least 15 morphologically similar species.

After five decades of intensive investigation, the classification of the algae is still a work in progress, with many taxa requiring far more research. There are many instances where a formal classification at the level of phylum or class is not designated and the algal group is referred to informally. For example, there is little agreement as to whether the close relatives of the land plants should be recognized as the phylum Charophyta or the class Charophyceae, meaning these organisms are often referred to as an informal group—the charophytes or charophyceans. This dilemma also needs resolving in the phylum Ochrophyta, which has 18 classes. Many of these had previously been recognized at phylum level, but are now widely recognized informally, reflecting the confusion.

CRYPTIC SPECIES

An area that has been yielding amazing results in describing algal biodiversity is the discovery of not only undescribed species in poorly surveyed habitats and geographical regions but also of cryptic species in well-known floras. Cryptic species are seemingly one species, often morphologically indistinguishable, but in reality they are two or more and sometimes many species masquerading as one. There are many examples of cryptic algal species. This hidden diversity was revealed in the brown seaweed *Lobophora variegata*, which had been regarded for 200 years as a single species widely distributed in tropical to warm temperate seas worldwide. Recent DNA studies have revealed that at least 15 species from different geographical regions had been recognized as this one species. DNA sequencing provides a time-efficient method that accelerates the rate at which new species are being discovered. Once discovered, the new species can be further circumscribed by morphological, biochemical, and other characters.

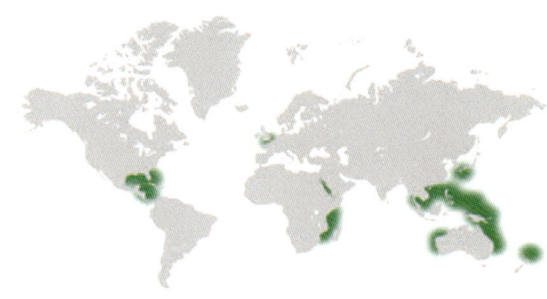

Lyngbya majuscula

Mermaid's hair

KINGDOM	:	Eubacteria
PHYLUM	:	Cyanobacteria
CLASS	:	Cyanophyceae
ORDER	:	Oscillatoriales
GENUS	:	*Lyngbya*
SIZE	:	Filament diameter 40 microns
HABITAT	:	A marine species of estuaries, sheltered bays, and coral reefs

Cyanobacterial species are usually microscopic unicellular, colonial, or filamentous organisms that can become visible to the unaided eye when they form large, luxuriant growths. Their cells contain chlorophyll *a* that is masked by phycobilin pigments—the blue-green phycocyanin and the red phycoerythrin.

The Cyanobacteria have two unique features: they are the only bacterial organisms included in the algae, and they are the only group of bacteria that photosynthesize and release oxygen.

Approximately 5,000 cyanobacterial species are estimated to exist on Earth. They occur most frequently as minor components in most marine, freshwater, and terrestrial habitats. Cyanobacteria often become ecologically dominant in harsh environments that are similar to those in which they evolved billions of years ago, or in those that have been damaged by human activities.

Lyngbya majuscula is a filamentous species, with each microscopic unbranched filament composed of a row of cells enclosed in a mucilage sheath. The species is a common inhabitant of estuarine mudflats and seagrass communities worldwide, particularly in tropical to warm temperate seas, although it also occurs on some coral reefs and in other marine habitats. It grows as a small group of filaments, as a mat covering the mud, or as a loosely entangled mass of filaments that are attached onto seagrass leaves. In shaded habitats the filaments are often bright red, due to the red photosynthetic pigment phycoerythrin, while in sunlit seagrass communities, the filaments of *Lyngbya majuscula* are brown, as the light-sensitive red pigment is degraded by the sunlight. Luxuriant growths of *Lyngbya majuscula* billow from seagrasses in long tresses that are reminiscent of human hair, hence its common name, "mermaid's hair."

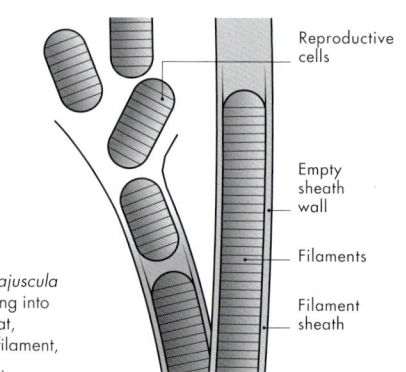

Reproduction in *Lyngbya majuscula*

Filaments of *Lyngbya majuscula* reproduce by fragmenting into short sections of cells that, once free of the parent filament, grow into new filaments.

Reproductive cells

Empty sheath wall

Filaments

Filament sheath

→ Light microscope image of the entangled filaments of *Lyngbya majuscula*. Enclosed in a colorless mucilaginous sheath, the filaments are composed of very narrow cells, whose longer dimension stretches across the filament. Many empty mucilaginous sheaths are visible in the background of the image.

KINGDOM	:	Plantae
PHYLUM	:	Rhodophyta
CLASS	:	Florideophyceae
ORDER	:	Ceramiales
GENUS	:	*Delesseria*
SIZE	:	Individual blades 2–10 in (5–25 cm) long, 1.2–4 in (3–10 cm) wide
HABITAT	:	A marine species of rocky subtidal habitats

PHYLUM RHODOPHYTA

Delesseria sanguinea

Sea beech

The phylum Rhodophyta (red algae) is a large, diverse group of approximately 7,000 known species, of which 95 percent are marine macroalgae (seaweeds). Red seaweeds are typically small- to medium-sized plants that are less than 12 in (30 cm) tall, although a few species grow much larger.

The red seaweeds grow in intertidal to deep subtidal habitats. The remaining 5 percent of Rhodophyta are either unicellular species, freshwater macroalgal species that live primarily in rivers and streams, or inhabitants of atypical environments, such as hot springs, soils, caves, and even sloth hair.

The Rhodophyta are classified in the kingdom Plantae based on the presence of the photosynthetic pigment chlorophyll *a* and the two-membrane plastid envelope. The Rhodophyta are also defined by the presence of floridean starch (the storage carbohydrate found only in the red algae); phycobilin photosynthetic pigments (particularly the red phycoerythrin); a cell wall of three carbohydrates (cellulose and the mucilaginous agar, or carrageenan); and the lack of flagellated cells. Agar and carrageenan, which are only found in the cell walls of the red seaweeds, have a multitude of uses in the food industry and medicine. Several species of red algae are farmed for their agar and carrageenan.

The elegant sea beech (*Delesseria sanguinea*) is a common perennial species that inhabits subtidal rocks on the Atlantic coasts of Europe. The species grows in the deep shade of lower intertidal rock pools, or more commonly as an understory species under the canopy of kelp forests and in open rock communities to depths of 100 ft (30 m). The plants attach to the rocks with a strong, disklike structure (holdfast), from which arise one to many stalked, flat, leaflike structures called blades. The blades are narrow to broad and taper to the blade tip; they have a prominent midrib from which arise visible lateral veins. The blade is delicate, membranous, and only one cell thick except for the well-developed tough midrib that is many cells thick. The blade margins are often ruffled. The plants of sea beech vary from blood red, as indicated by the species name, to rose in color. Victorian ladies favored this beautiful species for their seaweed albums.

→ A beautiful red seaweed. The delicate membranous blades of the sea beech are wrinkled, have ruffled margins, and are one cell thick except for the prominent midrib and lateral veins.

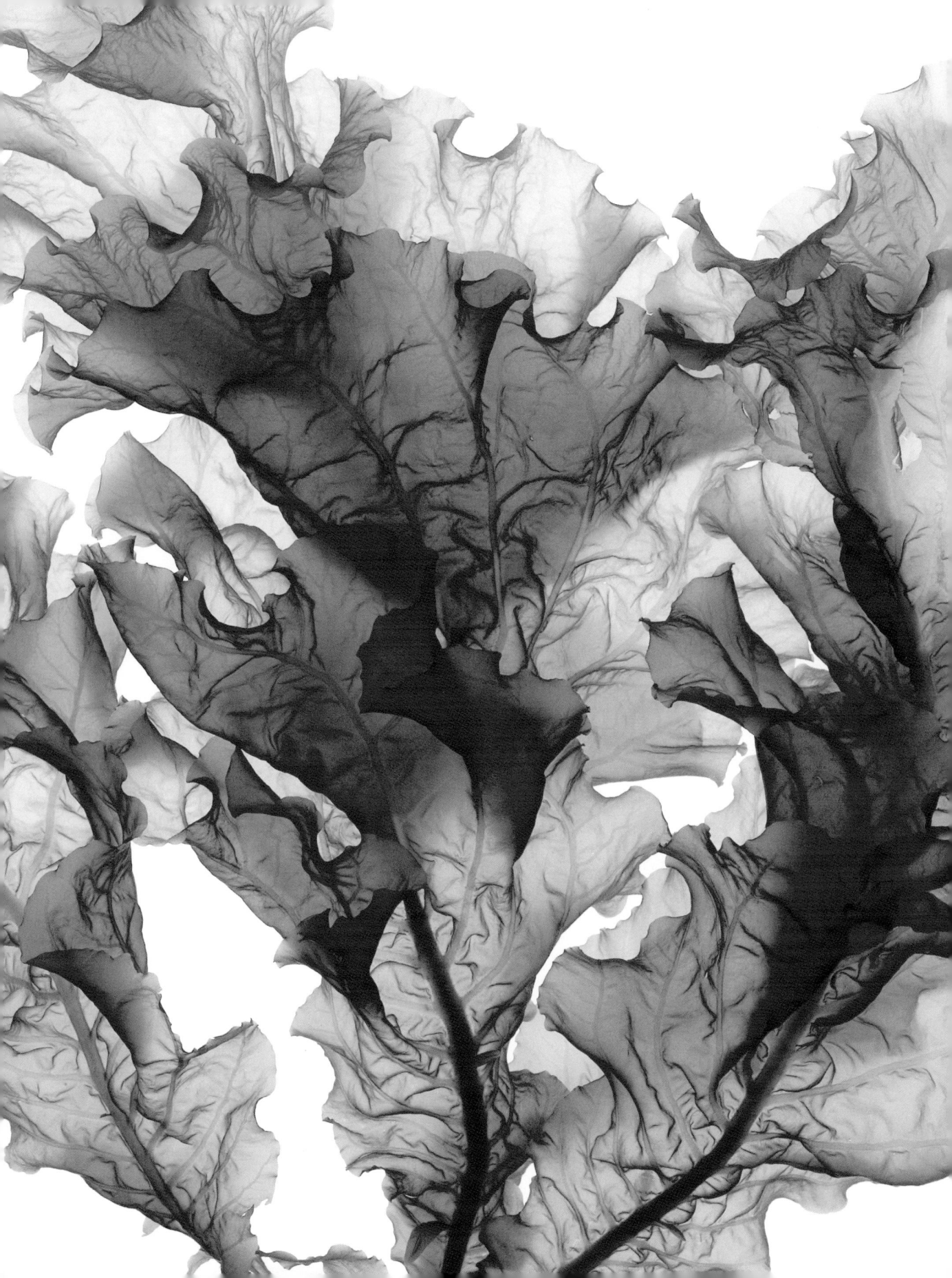

PHYLUM CHLOROPHYTA

Anadyomene lacerata

Green alga

KINGDOM	Plantae
PHYLUM	Chlorophyta
CLASS	Ulvophyceae
ORDER	Cladophorales
GENUS	Anadyomene
SIZE	Plants to 2.4 in (6 cm) tall
HABITAT	A deep sea species restricted to depths of 49–197 ft (15–60 m)

The phylum Chlorophyta (green algae) include unicellular, colonial, and macroalgal species that inhabit freshwater, marine, and terrestrial environments. Freshwater species live in lakes, rivers, streams, ponds, bogs, wetlands, and even garden birdbaths, while marine species inhabit rocky shores, estuaries, bays, coral reefs, and the world's oceans.

Still other green algal species have moved from aquatic to terrestrial environments, where they have adapted to the rigors of life on soils, tree trunks, leaves, fences, and snow. A very diverse, species-rich group, the Chlorophyta comprise approximately 8,000 described species, the majority of which live in freshwater habitats rather than in marine and terrestrial habitats. Interestingly, the main lineages in the Chlorophyta gave rise to either freshwater or marine species, unlike the Rhodophyta and Phaeophyceae, which consist predominantly of seaweeds.

Although red and green algae are both classified in the kingdom Plantae and possess plastids with a two-membrane envelope, the green algae have diverged more widely from their cyanobacterial ancestor. The green algae have lost the phycobilin pigments and acquired chlorophyll *b*. They have developed a more complex plastid with stacks of two to six to sometimes many thylakoid membranes and have starch as their storage carbohydrate.

The stiff, erect, shield-shaped thalli of *Anadyomene lacerata* grow as scattered plants attached by numerous thick-walled rhizoids to pebbles and rubble on subtidal sand plains. The thallus is composed of one layer of large multinucleate cells, which are organized into a beautiful lacy pattern of veins, fanned branches, and smaller cells. Veins are composed of one to three long, tubular, inflated club-shaped cells that radiate from the base to the upper edges of the plant body. In the middle of the thallus, these long cells can attain lengths of 0.14 in (3.44 mm) and widths of 0.01 in (0.31 mm). Each vein cell has, at its summit, two to six large cells that form fingerlike fanned branches. The spaces between the veins are filled with smaller cells that, being perpendicular to the veins, have a featherlike branching pattern.

→ The elaborate lacy thallus of the green seaweed *Anadyomene lacerata* is constructed from one layer of large cells organized into fan-shaped patterns of veins and parallel cells infilling the spaces between the veins.

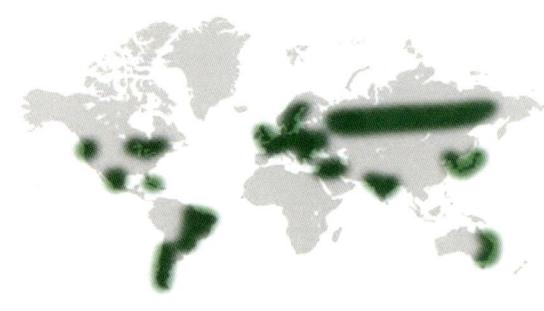

KINGDOM	:	Protozoa
PHYLUM	:	Euglenozoa
CLASS	:	Euglenophyceae
ORDER	:	Euglenida
GENUS	:	*Euglena*
SIZE	:	83–163 microns long, 10–24 microns wide
HABITAT	:	Fresh and brackish waters

PHYLUM EUGLENOZOA

Euglena deses

Unicellular alga

Historically claimed by botanists as plants and by zoologists as animals, euglenoids are elongate, spindle-shaped, spherical, cylindrical, flattened, or twisted unicells that most commonly bear two flagella, inserted into the anterior pocket in the cell. They are classified in the unicellular kingdom Protozoa.

In comparison to other unicellular algae, some *Euglena* species have large cells, reaching lengths of 200–500 microns. Approximately 2,000 euglenoid species have been described, the majority of which live in freshwater and brackish habitats, although a few species are members of the marine phytoplankton. Curiously, some *Euglena* species live among wet sand grains and color the sand green in areas of freshwater seepage on ocean beaches.

Around one half of euglenoid species lack plastids and are colorless, while the other half typically have many green plastids (known as chloroplasts) in their cells. This affects their mode of nutrition. The colorless euglenoids feed by absorbing dissolved organic compounds or by ingesting bacteria and unicellular prey. Photosynthetic euglenoids acquired their chloroplasts from a green algal cell through secondary endosymbiosis, which is evident from the three-membrane envelope bounding the chlorophyll *a*- and *b*-containing chloroplasts. The storage carbohydrate paramylon is unique to the euglenoids, as is the pellicle lying under the cell membrane.

The pellicle, which can be rigid or flexible, is composed of parallel interlocking protein strips wound helically around the cell. In species with a flexible pellicle, the cell changes shape when the protein strips flex and slide relative to each another. This causes a cytoplasmic bulge to flow from the posterior to the anterior end of the cell, propelling the cell backward. These curious squirming backward movements—often referred to as euglenoid movements—occur when the cells are not using their flagella to swim.

Cells of *Euglena deses* contain numerous lens-shaped chloroplasts and a large red-orange eyespot that directs the cell to swim toward light; its flexible pellicle permits a violent, twisting, and continuously turning snakelike movement. The species has one short flagellum that is often retracted into the anterior pocket. *Euglena deses* is widely distributed in freshwater and brackish puddles, swamps, ponds, lakes, and bogs, where it is an indicator of moderate to heavy organic pollution.

→   Light microscope image of *Euglena deses*. The numerous snakelike cells, with many chloroplasts and a red-orange eyespot, are easily distinguished from the brown cell of another euglenoid species, *Lepocinclis spirogyroides*, which has rows of shining beadlike projections on its cell surface.

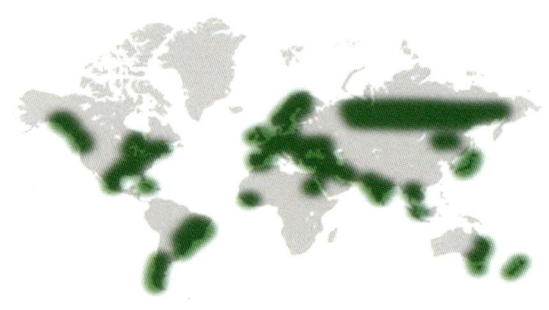

Ceratium hirundinella

Dinoflagellate

KINGDOM	Chromista
PHYLUM	Miozoa
CLASS	Dinophyceae
ORDER	Gonyaulacales
GENUS	*Ceratium*
SIZE	Cells 40–450 microns long
HABITAT	Freshwater lakes and ponds

Dinoflagellates are microscopic unicellular or colonial flagellates. Only half of the 2,500 described species possess plastids, which usually contain the red pigment peridinin and are bounded by a three-membrane envelope. Dinoflagellate plastids usually are derived from a red alga.

The plastid-bearing dinoflagellates are capable of photosynthesis, and may have either a phototrophic or a mixotrophic mode of nutrition (some mixotrophic dinoflagellates are osmotrophs, which absorb organic compounds from the surrounding water, while others are predators on other unicellular organisms).

Dinoflagellates are common to abundant in marine, estuarine, and freshwater environments, although only 10 percent— approximately 250 species—occur in freshwater habitats. Although they range in length from 2 to 2000 microns, the majority are less than 100 microns long, but they all share a unique cell covering, flagellar orientation, and nuclear organization. The cells are covered by flattened thecal vesicles lying under the cell membrane, which appear empty in unarmored species or contain thecal plates (the armor) in armored species. Dinoflagellate cells have two flagella, one in a girdle encircling the cell and the other in the longitudinal groove or sulcus. In the nucleus of the dinoflagellates, the chromosomes are visible in nondividing cells, unlike in other eukaryotes in which the chromosomes are only visible during cell division.

The broad to narrow spindle-shaped cells of *Ceratium hirundinella*, which measure from 40 to 450 microns in length, are large for a dinoflagellate species. The species is easily recognized by its distinctive morphology. The epitheca (the anterior half of the cell) is drawn out into a single hollow apical horn, while the hypotheca (the posterior half) bears two or three hollow posterior horns, one horn being considerably longer than the other one or two. The cell is armored, covered by relatively thick thecal plates. This mixotrophic species is common in nutrient poor to nutrient enriched freshwater ponds and lakes.

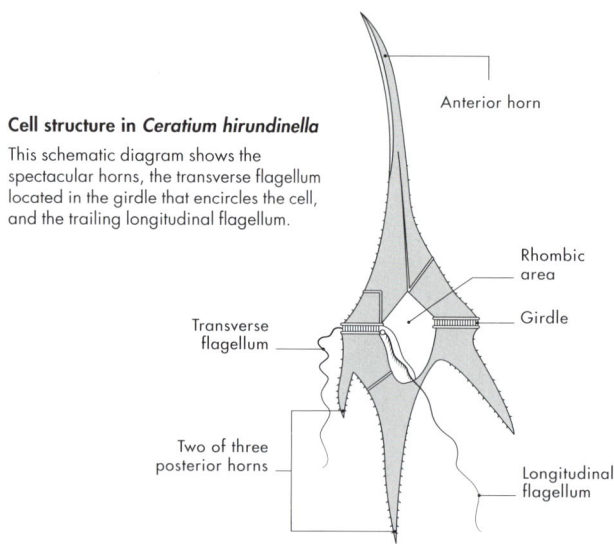

Cell structure in *Ceratium hirundinella*
This schematic diagram shows the spectacular horns, the transverse flagellum located in the girdle that encircles the cell, and the trailing longitudinal flagellum.

Anterior horn

Rhombic area

Girdle

Transverse flagellum

Two of three posterior horns

Longitudinal flagellum

→   The light microscope image of the distinctive cell of *Ceratium hirundinella*.

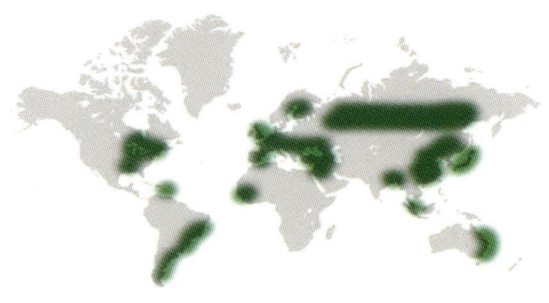

KINGDOM	:	Chromista
PHYLUM	:	Cryptophyta
CLASS	:	Cryptophyceae
ORDER	:	Cryptomonadales
GENUS	:	*Cryptomonas*
SIZE	:	Cells 14–80 microns long, 5–20 microns wide
HABITAT	:	Freshwater planktonic species

PHYLUM CRYPTOPHYTA

Cryptomonas ovata

Unicellular alga

The phylum Cryptophyta are microscopic unicellular flagellates that can be colorless, blue-green, reddish, red-brown, olive-green, or brown. The photosynthetic cryptophytes have retained the blue-green and red phycobilin pigments of their red algal ancestor.

The presence of the phycobilin pigments distinguishes the cryptophytes from the other algal groups that have also acquired their plastids from a red alga but have lost these pigments. The cryptophytes are also characterized by their cell covering (the periplast), which is a series of proteinaceous plates below the cell membrane, and an anteriorly located furrow–gullet system from which two hairy flagella emerge.

Around 200 species of cryptophytes have been described from marine, estuarine, and freshwater habitats. They are important components of phytoplankton communities worldwide and often increase in abundance when other phytoplankton groups are declining, particularly during periods of environmental disturbance. Pulses of cryptophytes commonly occur following the environmental disturbances caused by major rain events.

Cryptophytes are often prolific in aquatic ecosystems, and are common food organisms for many unicellular organisms. Containing high levels of polyunsaturated fatty acids (among other nutrients), cryptophytes are generally regarded as a high-quality food for aquatic herbivorous unicellular species, including dinoflagellates and ciliates.

Cryptomonas ovata is a common species that inhabits the plankton of freshwater habitats, where it forms blooms. As the species name implies, the cells are oval in shape and have two plastids. *Cryptomonas ovata* supplements the energy it produces by photosynthesis by ingesting bacteria in low-light environments and thus has a mixotrophic mode of nutrition.

Cell structure of *Cryptomonas ovata*

The two flagella are inserted into the vestibulum, below the opening of a groove that runs longitudinally through the cell and terminates in the saclike gullet. The gullet is lined with longitudinal rows of ejectosomes, organelles that eject fine threads.

Flagellum

Vestibulum

Gullet

Nucleus

Ejectosome

Periplast (cell covering)

→ Rare light microscope images of *Cryptomonas* (Latin, meaning "hidden individual"). The oval cells (front view) have two brown plastids and two whiplike flagella, while the gullet lined with rows of ejectosomes is visible in the conical cells (side view).

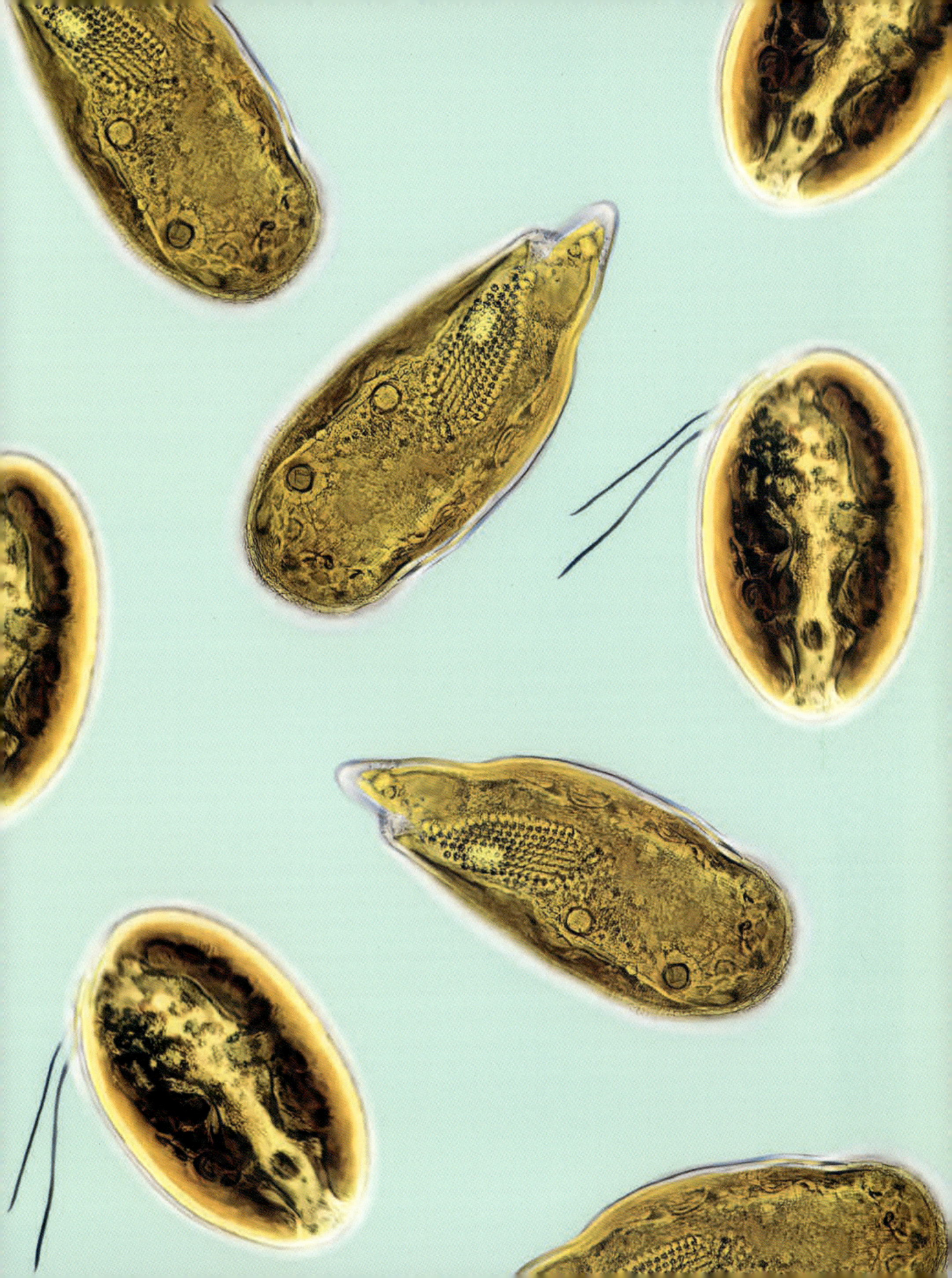

KINGDOM	Chromista
PHYLUM	Bacillariophyta
CLASS	Mediophyceae
ORDER	Thalassiosirales
GENUS	*Thalassiosira*
SIZE	Cells 10–60 microns in diameter
HABITAT	Planktonic in coastal seas

PHYLUM BACILLARIOPHYTA

Planktoniella sol

Centric diatom

The microscopic unicellular and colonial species of the phylum Bacillariophyta are the best-known and most abundant group of the planktonic algae. Commonly called diatoms, they are defined by their unique cell wall, the frustule, which is composed of silica and organic material.

The diatoms' boxlike frustule, which ranges in size from 2 to 200 microns, is exquisitely ornamented with patterns of pores, ribs, and spines that radiate in spectacular geometric symmetry across the surfaces of the cells.

The lack of flagella on the diatom vegetative cell is another character that distinguishes the diatoms from the majority of algal phyla. Species of diatoms are either planktonic and float in the water column or live attached to various surfaces in oceans, seas, rivers, lakes, and many other habitats. Although they lack flagella, some diatom species can glide relatively rapidly on surfaces onto which they have secreted a layer of mucilage. The male gametes of one group (centric diatoms) are the only flagellate cells in the diatoms. These male gametes have one hairy flagellum that is directed forward when swimming. They lack the smooth posterior flagellum typical of the other heterokont algae (algae that have one hairy flagellum and one smooth flagellum).

The number of diatom species occurring on Earth is not known, but they are nevertheless a species-rich group that has been wildly successful in terms of evolutionary diversification.

Estimates vary widely, with a current conservative estimate of 20,000 species, significantly less than 100,000 species proposed by some diatom specialists. Yet regardless of the actual number, diatoms are important primary producers in aquatic ecosystems worldwide, underpinning the food webs that support highly productive natural ecosystems and commercial fisheries.

The pillbox-shaped cell of *Planktoniella sol* sits in the center of a prominent flaplike membranous wing, a structure that is uncommon in the diatoms. This winglike ribbed extension of organic material is extruded from and remains attached to the side of the cell. It is thought to function as a flotation device aimed at keeping the cells of *Planktoniella* in the sunlit surface waters of the sea. Diatoms that lack flagella, and therefore cannot swim, possess a variety of structures that maintain the buoyancy of their cells. The boxlike frustule covering the diatom cell consists of two valves that are held together by girdle bands, with the smaller valve fitting into the larger one. Diatom cells look different when viewed from the top or bottom (valve view) compared to when viewed from their sides (girdle view). The top or bottom view of the *Planktoniella* frustule reveals an elegantly ornamented valve face covered with rows of pore-like depressions radiating from the center, whereas in the side view the rows of smaller pores are separated by girdle bands that lack ornamentation.

→ Electron microscope image of the pillbox-shaped cell of *Planktoniella sol* surrounded by the organic wing.

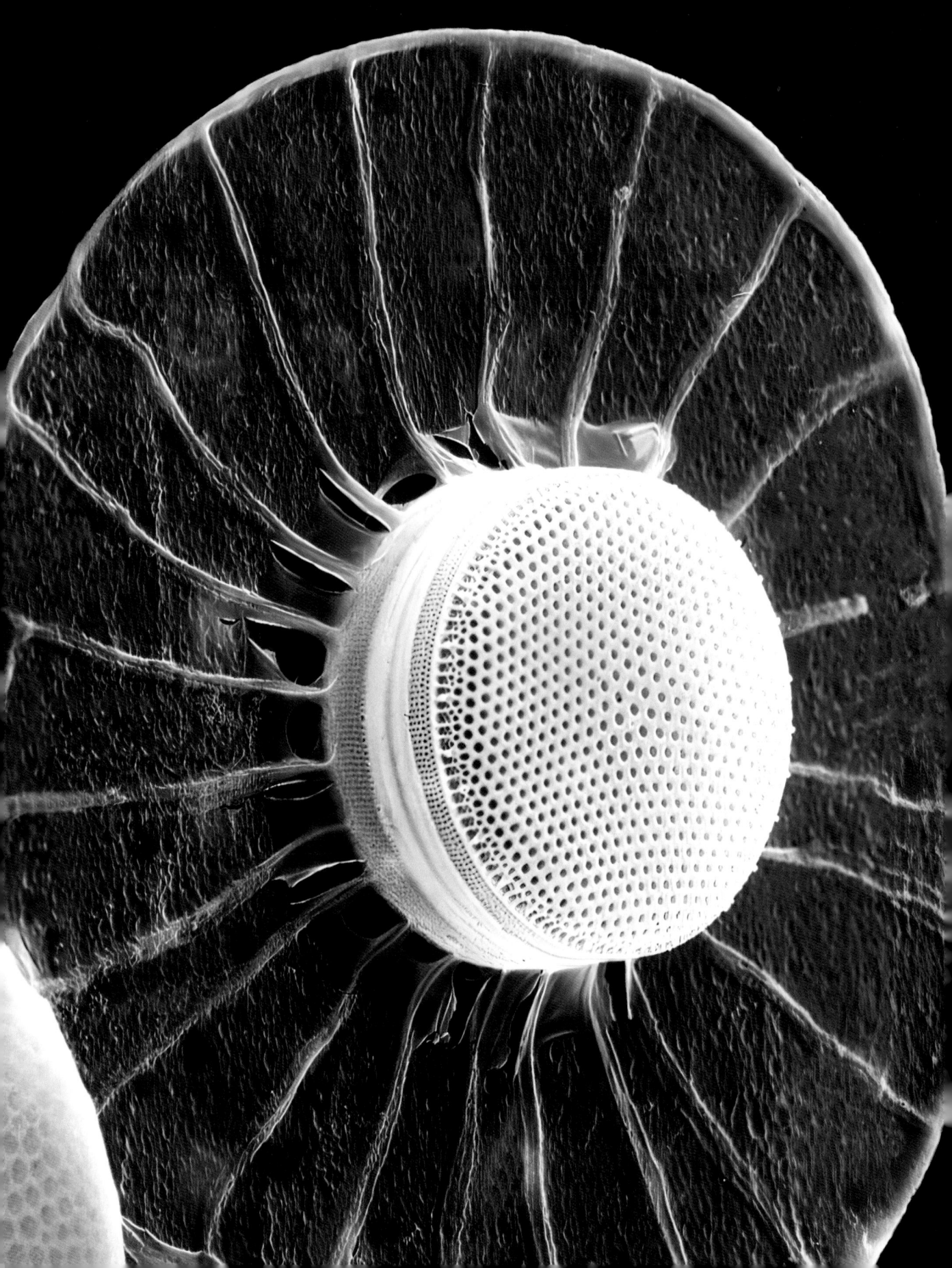

Fucus vesiculosus

Bladder wrack

KINGDOM	Chromista
PHYLUM	Ochrophyta
CLASS	Phaeophyceae
ORDER	Fucales
GENUS	*Fucus*
SIZE	Blades usually 7.9–27.5 in (20–70 cm) long, 0.2–1.6 in (0.5–4 cm) wide
HABITAT	Mid intertidal zone on seashores

The class Phaeophyceae (brown algae) is loosely assigned with 17 other classes to the phylum Ochrophyta, based on their dissimilar flagella comprising a hairy anterior flagellum and a smooth posterior flagellum.

The brown seaweeds are distinguished from other algae by the possession of chlorophyll *a* and *c*; a cell wall composed of cellulose, alginic acid, and fucoidan; the storage carbohydrate laminarian; and their complex multicellularity. Approximately 1,792 species of the brown algae have been described worldwide, most of which are marine species, although a few species have invaded freshwater habitats from the sea. They range in appearance and size from microscopic filaments to large leathery kelps up to 100 ft (30 m) long. There are no unicellular brown algal species. Carpeting the rocky intertidal shores in cool temperate regions and forming complex subtidal forests from the tropics to polar regions, the brown seaweeds function as ecosystem engineers, providing a habitat and food for diverse assemblages of other algae and marine animals.

On European shores, species of *Fucus* are common intertidal seaweeds that often form wide bands across the rocks in the upper intertidal (*Fucus spiralis*), mid-intertidal (*Fucus vesiculosus*), and lower intertidal (*Fucus serratus*) zones.

Fucus vesiculosus derives its name from the vesicles or air bladders that float thalli into the sunlit waters of the sea. The flat, leathery, straplike thallus of this species repeatedly branches into two equal branches. Each branch has pairs of bladders, which are generally (but not always) present either side of the prominent midrib, except at the branching point where there is only one. The bladders are almost spherical, protruding equally on both sides of the thallus surface.

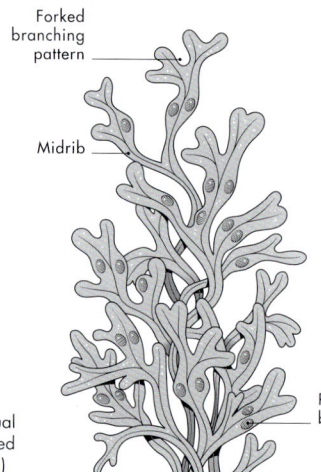

Forked branching pattern

Midrib

Paired air bladders

Thallus structure of the bladder wrack

The thallus branches into two equal branches and generally has paired vesicles (air bladders for flotation) either side of the conspicuous midrib.

→ The leathery, flat, straplike thallus of the bladder wrack is distinguished from other brown seaweeds by its forked branching pattern, the presence of a prominent midrib, and usually paired vesicles.

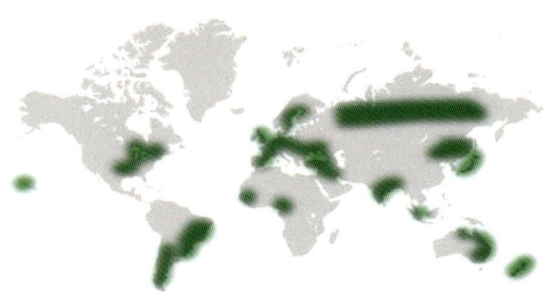

CLASS CHRYSOPHYCEAE

Dinobryon sertularia

Golden-brown alga

KINGDOM	Chromista
PHYLUM	Ochrophyta
CLASS	Chrysophyceae
ORDER	Chromulinales
GENUS	*Dinobryon*
SIZE	Individual lorica 30–40 microns long, 10–14 microns wide
HABITAT	Planktonic in freshwater habitats

The class Chrysophyceae (golden-brown algae) range in form from unicellular flagellates and colonies to filaments. Around 430 chrysophycean species have been described, most of which occur in freshwater habitats, with a few species found in estuarine and marine waters.

The Chrysophyceae are one of the 18 classes of the phylum Ochrophyta of the much larger group, the heterokont algae. The chrysophyceans contain the pigments chlorophyll *a* and *c*, and the brown pigment fucoxanthin, as well as possessing two dissimilar or heterokont flagella (an anterior hairy flagellum

and a posterior smooth flagellum). However, unlike most other heterokonts, cells of these golden–brown algae are either naked (lacking cell walls) or covered by a silica exoskeleton, silica scales, or a cellulose lorica (a protective covering surrounding the cell).

Species of *Dinobryon* are either solitary or colonial, with cells surrounded by a cellulose, vaselike lorica. *Dinobryon sertularia* is a colonial species that forms dense, branched, treelike colonies. These grow when a newly divided cell settles and builds a lorica on the edge of an existing lorica. Inside each lorica, a thin cytoplasmic thread attaches the cells to the lorica's base. The cells have two flagella (the longer flagellum extends into the water beyond its lorica), two plastids, and, at the cell's apex, a light–sensitive red eyespot that directs the cell to swim toward light. Each cell's flagella beat, enabling the colony to swim in a graceful—albeit rather slow—manner. Although it is photosynthetic, *Dinobryon sertularia* also engages in substantial rates of bacterivory. This is an adaptation to life in slightly alkaline, relatively nutrient-poor lakes and pools. This strategy, which supplements the uptake of carbon and other nutrients, enables it to maintain its abundance in the plankton during periods where poor light and nutrient conditions are less suited to growth.

A colony of *Dinobyron sertularia*

Individual cells, each in the vase-shaped lorica (A), form the colony (B) by dividing into two to four cells which either remain in the parental lorica or produce their own lorica at the top of the parental lorica.

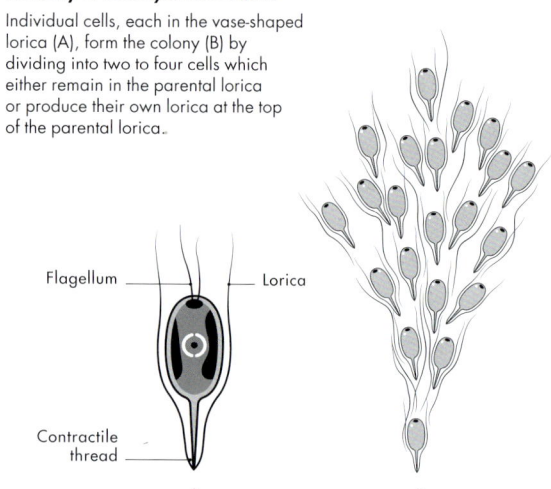

Flagellum — Lorica

Contractile thread

A   B

→ A swimming colony of the planktonic golden-brown alga *Dinobryon sertularia* is composed of many flagellate cells, each located in a translucent cellulose vase-shaped protective covering (lorica).

MORPHOLOGY

The megadiverse algae

Algae are curious living organisms that exist in every conceivable shape and size. The algal body—called the thallus—ranges from the relative simplicity of a microscopic single cell measuring just 1 micron in diameter, to the complexity of giant leathery kelps that can grow more than 160 ft (50 m) in length. The external form and structure—termed morphology— are important characteristics used to describe algal species.

AN AMAZING ARRAY OF FORMS

The morphological diversity encompassed by the algae is enormous. Single algal cells are not just simple red, green, or yellow-brown spheres; they have evolved into the shapes of spindles, crescents, pyramids, stars, cubes, cones, and even more complicated configurations, with their cell margins bearing horns, lobes, ridges, spines, and winglike extensions. One group of freshwater green algae, the desmids (order Desmidiales), are remarkable for the amazing array of their morphology, the often extraordinary complexity of their cell outlines, and their marked symmetry.

The seaweeds are equally diverse, with thalli (algal bodies) that form delicate or robust filaments, sheets, tubes, blades, fronds, crusts, sacs, cushions, and large, leathery straps. Yet despite this enormous morphological variation, there are five key, basic body plans that are used to describe the different levels of thallus organization in the various lineages. These are unicellular, colonial, multicellular, siphonocladous, and siphonous.

The body plan of a unicellular species is a single cell. In the colonial body plan, many individual cells of similar form and function are loosely held together by a viscous secretion known as mucilage, or interlocking structures or by other means. The multicellular body plan is an advanced level of organization, with numerous closely adherent cells that have the capacity to communicate with each other and to specialize. In siphonocladous body plans,

thalli are composed of large cells with numerous nuclei (called multinucleate cells). The siphonous multinucleate body plan takes the form of a long tube—the siphon.

Having evolved independently, these five basic body plans are found in the various algal lineages. For example, all five body plans are identifiable among species of the very diverse green algae (Chlorophyta) but only the multicellular body plan has evolved in brown algae (Phaeophyceae), almost all of which are seaweeds or marine macroalgae (algae that are visible to the unaided eye). Microscopic unicells and colonies are referred to as microalgae.

Although there are only five basic body plans, there are still plenty of variations in morphology within them. Unicellular and colonial species differ markedly in cell structure, and may be motile—capable of movement— or nonmotile—incapable of movement.

→ Phycologist David Williamson's poster collage illustrates a small proportion of desmid morphological diversity, which includes the multilobed circular cell of *Micrasterias* (top left corner).

MULTICELLULAR BODY PLAN

Multicellular thalli are composed of either filaments that consist of a row of cells joined end to end or tissues formed from three-dimensional arrays of cells. Protoplasts of these cells are interconnected to the adjoining cells by strands of cytoplasm. Some multicellular thalli that superficially appear to be composed of tissues are actually formed from tightly adherent branched filaments, a thallus construction that occurs in the majority of red seaweeds.

Most macroalgae, including *Laurencia brongniartii*, show no differentiation into specialized structures: all parts of the thallus look similar. However, some large brown seaweeds, particularly the brown algal genus *Sargassum*, display a level of body differentiation that is similar to the flowering plants, with rootlike holdfasts, stemlike stipes, and primary axes, leaflike lateral branches, and reproductive structures.

Macroalgae are anchored to rocks or other firm surfaces by a holdfast. Rhizoids, the rootlike filaments in the holdfast, glue the macroalgae onto the rocks. The macroalgal thallus usually grows from a meristem—a cell or a group of closely associated cells that are capable of repeated cell divisions. Meristems are located at the tip (apical meristem) or in the middle of the thallus (intercalary meristem).

Many algal species secrete complex extracellular carbohydrates in the form of gels, mucilages, and slimes, such as the mucilaginous adhesive that fastens the holdfast onto a rock. These have many important functions in the algae, although the presence of these extracellular substances has led some people to unkindly refer to algae as "slimy" or "green slime."

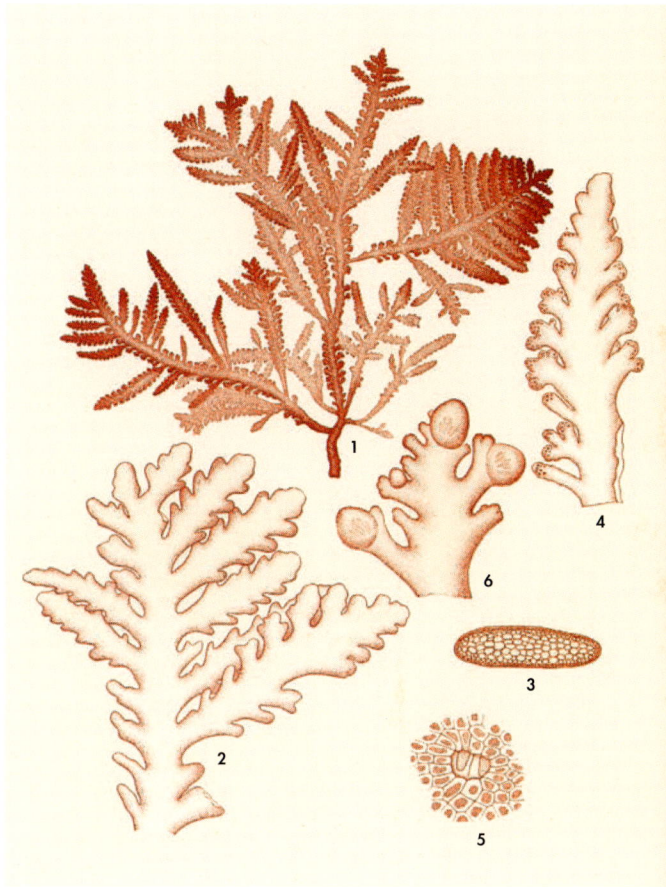

↑   All parts of the simple undifferentiated thallus of the red seaweed *Laurencia brongniartii* are similar in appearance.

←   The thallus of the brown seaweed *Sargassum platycarpum* is differentiated into stemlike primary branches that are attached to rocks by a rootlike holdfast (1), fertile branches (2), leaflike lateral branches (3), reproductive structures (4), and gas-filled pneumatocysts (5).

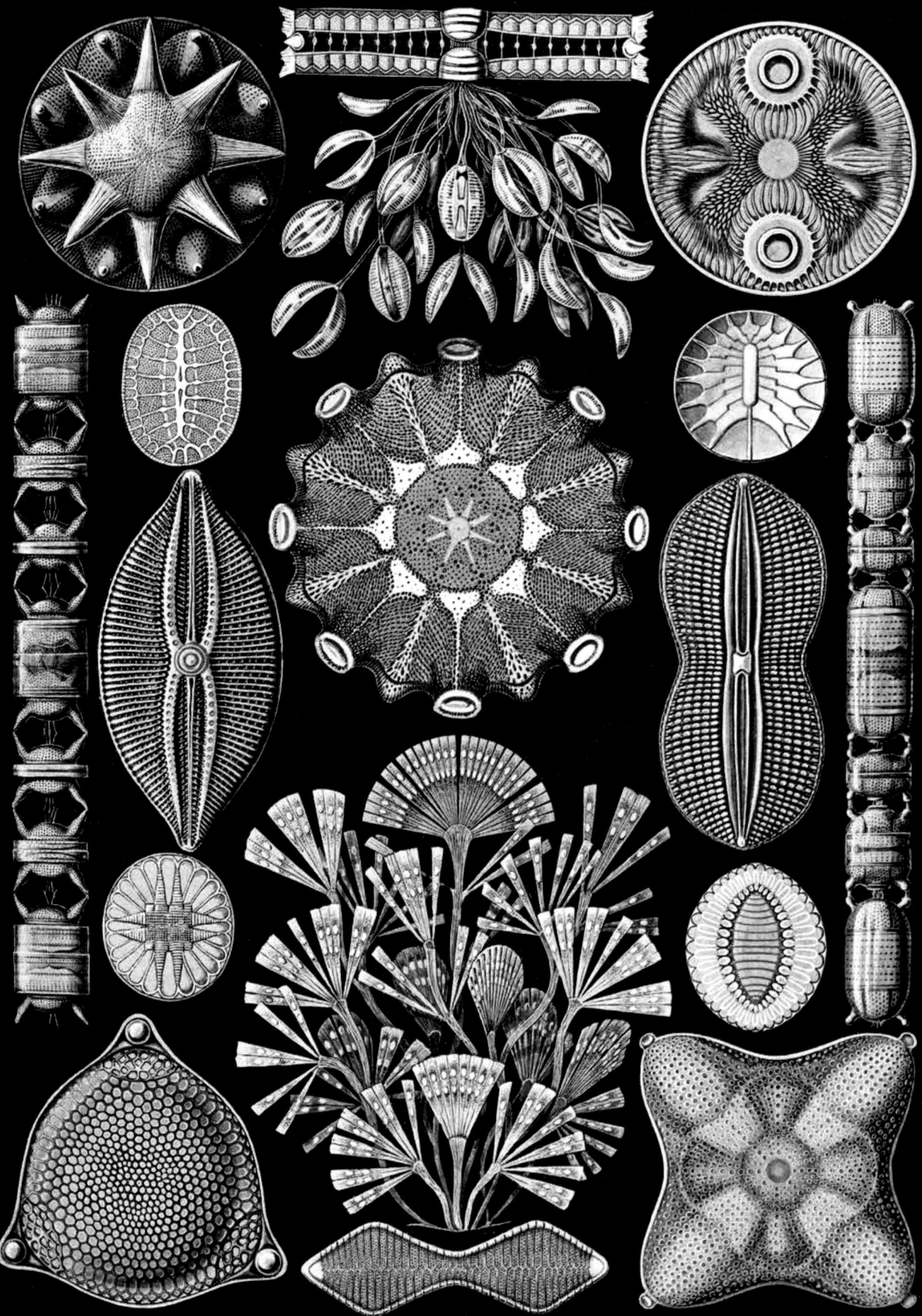

Unicells

Unicellular algae exhibit an elaborate variety in their single-cell structure, which differs markedly in each algal lineage. They can be nonmotile or motile, with motile species (called algal flagellates) bearing flagella—whiplike appendages that beat to propel the cell through the water.

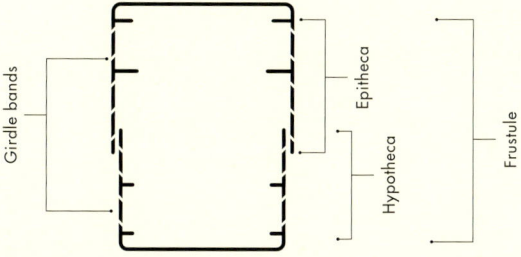

Life in glasshouses

The smaller posterior valve (hypotheca) of the boxlike frustule fits inside the larger anterior valve (epitheca), the two valves of the diatom cell held together by the girdle bands. Glass is made from silica.

← A page from Ernst Haeckel's *Diatomeae* (1904) illustrates a small proportion of the morphological diversity encompassed by unicellular and colonial centric and pennate diatoms, including the pennate colonial *Licomorpha* (middle, bottom row).

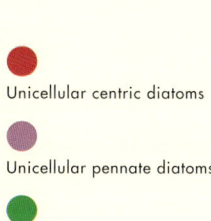

● Unicellular centric diatoms

● Unicellular pennate diatoms

● Colonial centric diatoms

● Colonial pennate diatoms (includes the mucilaginous stalks that the diatoms secrete)

THE DIATOMS

The most diverse of the phytoplankton phyla are the diatoms (phylum Bacillariophyta). Their cells range in size from 5 to 200 microns. Unique in the algae, their distinctive, rigid, boxlike silica cell wall (the frustule) is composed of two halves that are known as valves. The smaller valve (the hypotheca) fits into the larger valve (the epitheca), with one or more girdle bands holding the two together.

Exquisitely ornamented, the pores, slits, tubes, ribs, and spines that radiate in spectacular geometry across the cell's surface aid buoyancy, lightening the heavy frustule and increasing the contact between the cell's cytoplasm (the protoplasm excluding the nucleus) and the external environment.

The diatoms comprise two groups of centric diatoms with radial symmetry, and a third group of pennate diatoms with bilateral symmetry. The vegetative (or nonreproductive) cells of the diatoms lack flagella, although some diatom cells are motile. Some pennate diatoms secrete mucilage onto the surface and rapidly glide on it. The male gametes of the centric diatoms are the only flagellate cells in the diatoms. They have one flagellum instead of the two flagella that are typical of algal species.

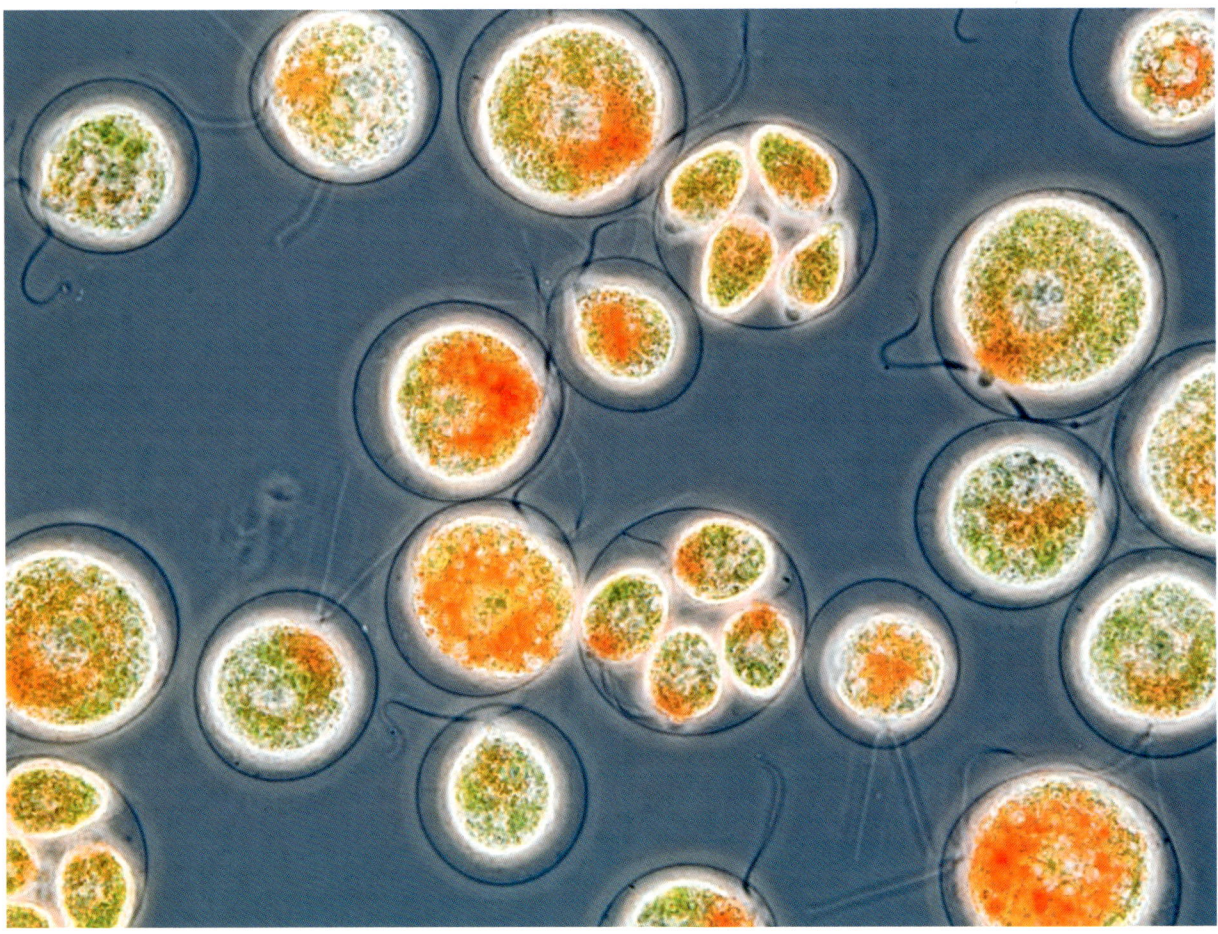

THE DESMIDS

The desmids (order Desmidiales) are a large group of freshwater green algae whose nonmotile unicells are renowned for their variety and the beauty of their shape and structure (see page 63). They are often much larger than other unicellular algae, as demonstrated by the cells of *Closterium acerosum*, which range in size from 300 to 600 microns in length and from 30 to 50 microns across. Species of *Closterium* can have simple, ellipsoidal, spindle-shaped, or cylindrical cells. The more intricate cells of *Micrasterias* are strongly compressed and almost circular (see page 63). They are divided into two halves by a prominent constriction and elaborately dissected into many leaflike or starlike lobes, with each half of the cell the mirror image of the other.

↑ The cell of *Haematococcus lacustris*, which is surrounded by a transparent extracellular matrix, has two whiplike flagella. Some cells have divided into four cells.

Desmids with cylindrical cells are similar when viewed from the front and side, but others appear different depending on the viewing angle. When the desmid *Staurastrum* is seen from its broad (front) side, each half-cell appears cup-shaped, with two long, armlike processes radiating at 45 degrees from each end of the two cups. However, in top view, the cell is triangular with three armlike processes; the fourth arm is either obscured by the cell or out of focus when the cell is viewed from its broadest side.

GREEN FLAGELLATES

There are numerous flagellate unicellular algal species. The green flagellates and the dinoflagellates are the best known and are both remarkably diverse. The green alga *Chlamydomonas* and its close relatives are flagellated unicells or colonies. The cells have two or four flagella and are usually covered by a thin glycoprotein layer that is external to the plasma membrane, which is thought to be "leaky." This cell covering would explain the unusual presence of two or more contractile vacuoles in the cytoplasm near the bases of the flagella in almost all species of the Volvocales—in unicellular organisms, the contractile vacuoles excrete excess water from the cell.

The cell covering of the biflagellate cells of *Haematococcus lacustris* is quite spectacular. These cells have radiating strands of cytoplasm that extend to the periphery of the wide and conspicuous watery and a sticky (mucilaginous) extracellular layer. In other volvocalean unicells the glycoprotein extracellular layer is less developed, and radiating cyptoplasmic strands are uncommon. *Haematococcus lacustris*, commonly known as the "blood-red alga," synthesizes and accumulates large amounts of carotenoids that are of great economic importance. There has been much study into the benefits of carotenoids in the prevention of a number of diseases in humans, including colon cancer.

The well-studied *Chlamydomonas reinhardtii* has been intensively researched by scientists for decades. The pear-shaped cell of this species measures from 20 to 30 microns in length and from 10 to 20 microns across. It is bounded by the plasma membrane, which is covered by a complex translucent three-layered extracellular matrix, with the outermost layer forming a highly organized capsule. The cell has a centrally located nucleus, many mitochondria, and—centered posteriorly in the cell—a large, cup-shaped chloroplast containing an eyespot. Two flagella are inserted at the cell's apex. The photosensitive eyespot apparatus directs the cell to swim toward light.

Flagella

Cell wall

Nucleus

Chloroplast membrane

Pyrenoid

Eyespot

Starch granule

Chloroplast (plastid)

Plasma membrane

Chlamydomonas reinhardtii

Considered the closest unicellular relative to and the basic unit of the colonies of *Gonium* and *Volvox*, cells of the green alga *Chlamydomonas reinhardtii* have two anterior flagella and a large, cup-shaped chloroplast with a pyrenoid (a starch storage body) and an eyespot.

THE DINOFLAGELLATES

The dinoflagellate cell usually consists of two parts: an anterior epicone and a posterior hypocone, which are separated by a groove (the cingulum) that encircles the cell. On the cell's ventral surface, a smaller groove (the sulcus) extends posteriorly from the cingulum and, in some species, anteriorly as well.

The two flagella on the dinoflagellate motile cells are also distinctive. They are inserted into the cell at the intersection of the two grooves. Both flagella are situated in the grooves. One flagellum—the transverse flagellum—is situated in the groove that encircles the cell, and the other flagellum—the longitudinal flagellum—is located in the smaller groove and trails behind the cell.

Dinoflagellate cells range in size from 5 to 200 microns, with some reaching 0.04 in (1 mm) in length (although this is rare). Another unique feature, a layer of flattened sacs (the thecal vesicles) lies under the plasma membrane of the dinoflagellate cell. These vesicles either appear empty or almost empty in the unarmored species, or they contain cellulose thecal plates of varying thickness in the armored species. The vesicles fit together tightly like jigsaw puzzle pieces

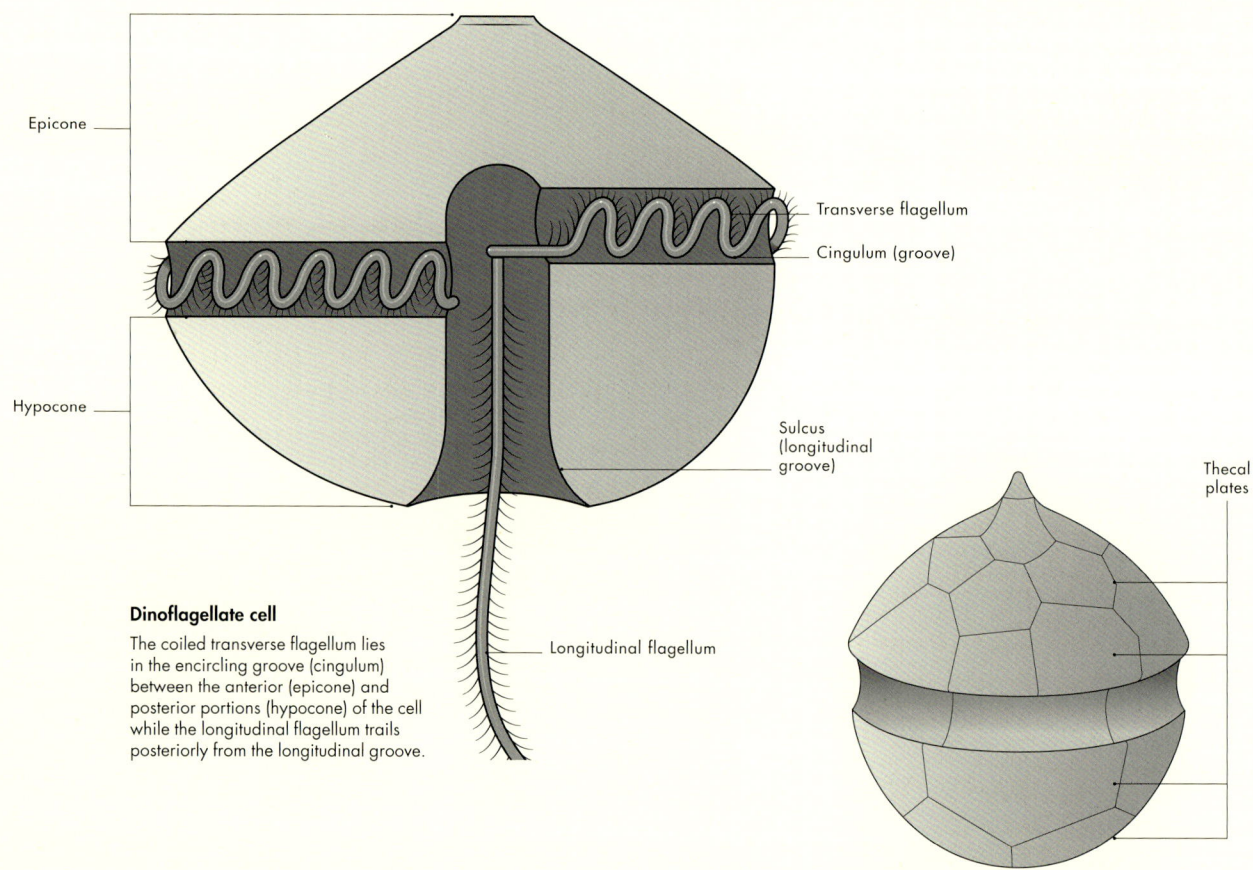

Epicone

Hypocone

Transverse flagellum

Cingulum (groove)

Sulcus (longitudinal groove)

Longitudinal flagellum

Dinoflagellate cell
The coiled transverse flagellum lies in the encircling groove (cingulum) between the anterior (epicone) and posterior portions (hypocone) of the cell while the longitudinal flagellum trails posteriorly from the longitudinal groove.

Thecal plates

Armored or not
Armored dinoflagellate cells are covered by cellulose plates that fit together like jigsaw puzzle pieces to completely cover the cell.

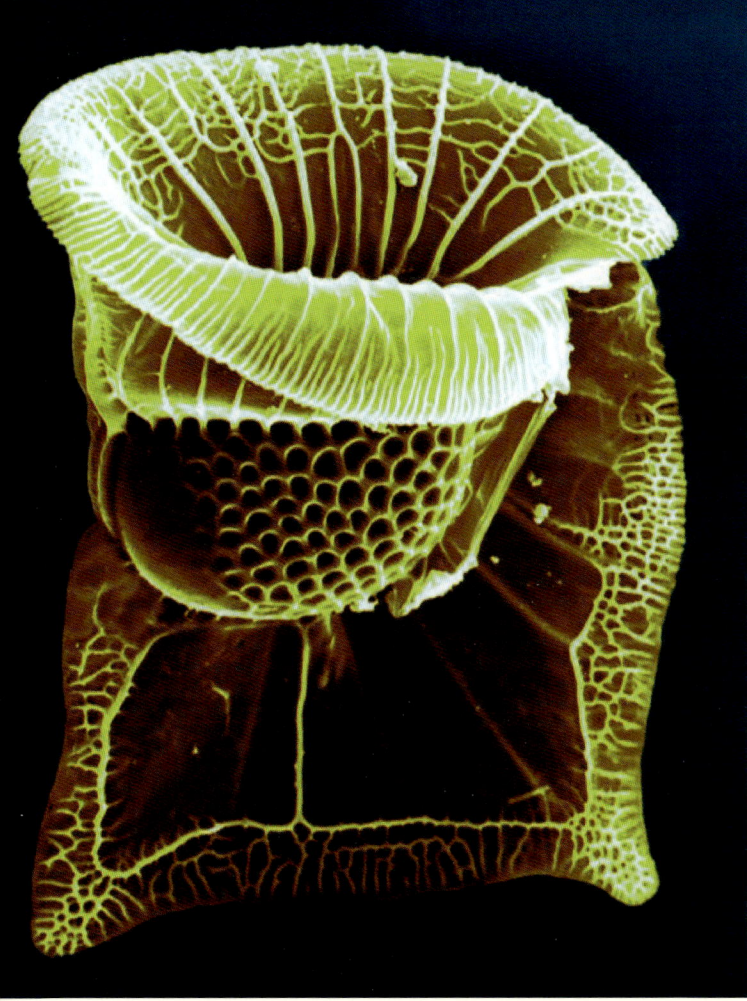

← A false-colored scanning electron micrograph of the magnificent dinoflagellate *Ornithocercus quadratus* shows the small spherical cell covered by pore-bearing thick thecal plates and overshadowed by spectacular large wings.

to encase the cell and may be heavily ornamented with pores, depressions, ridges, reticulations, and spines. The number, shape, and arrangement of the thecal vesicles in unarmored species and the cellulose plates in armored species differ in each dinoflagellate species and therefore are important for species identification.

Dinoflagellate morphology is astonishingly diverse. Species of *Tripos* and the closely related genus *Ceratium* (see page 50) have the typical dinoflagellate structure, except for the spectacular horns that define the genera. The hollow horns may have either open or closed pointed tips. The epitheca is drawn out into a single elongated apical horn, while the hypocone has two or three elongate horns. The number, size, and shape of the horns are important characters used to identify the approximately 100 species of *Tripos* and *Ceratium*.

Some tropical dinoflagellate species exhibit beautiful or strange morphological modifications to the spines, horns, and winglike structures more typical of the dinoflagellates. These morphological extensions, which increase the surface area to volume ratio, are thought to aid flotation and enhance nutrient uptake, as well as protecting against predation by small zooplankton. As the names suggest, *Ornithocercus magnificus*, *Ornithocercus splendidus*, and other species in the genus are magnificent. The small spherical cell of *Ornithocercus quadratus* is overshadowed by two funnel-shaped wings (lists), one located either side of the transverse groove and a third wing on the lateral and posterior cell surfaces. The wings are reinforced by a network of ribs.

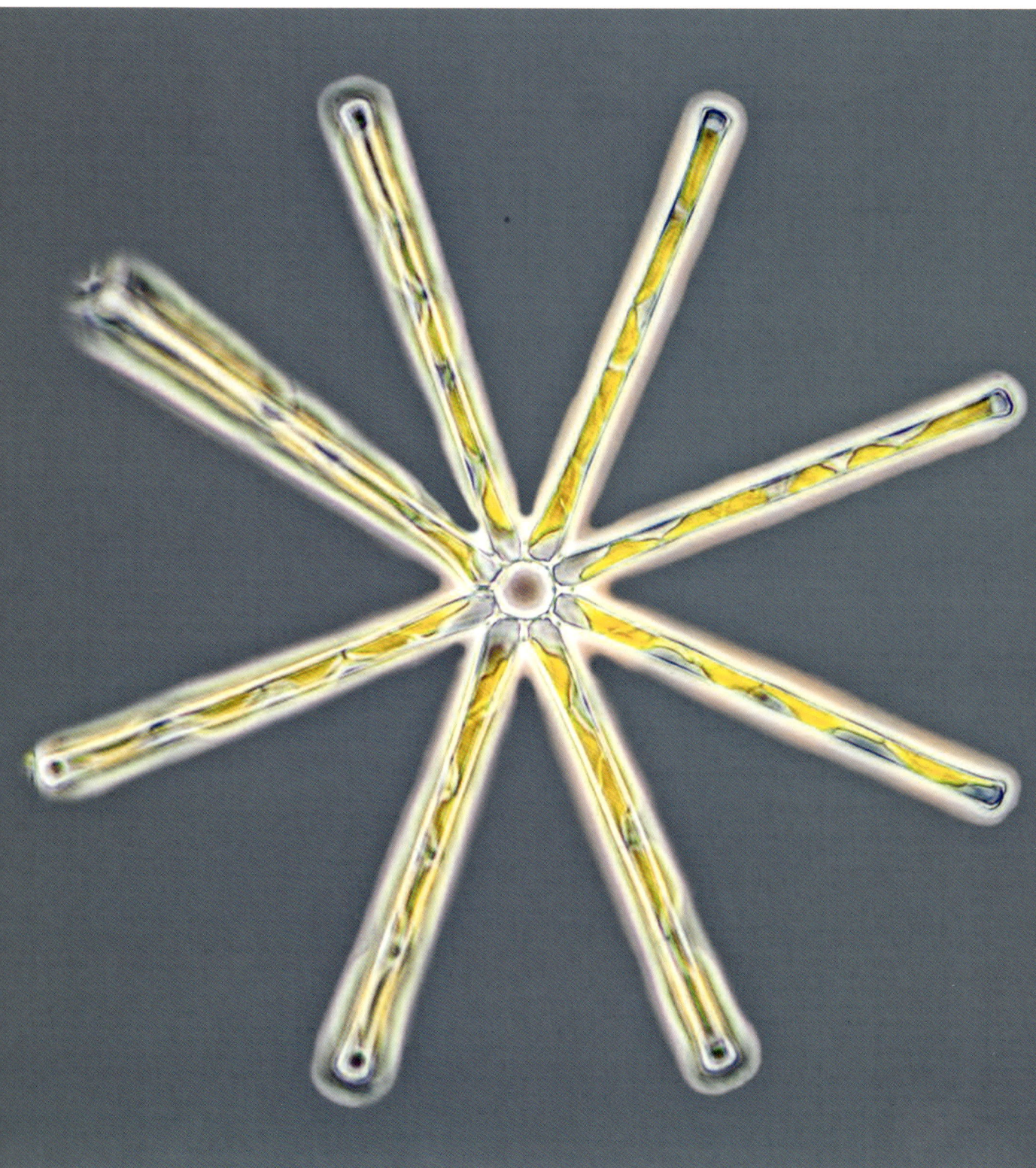

Colonies

The colonial body plan occurs across many algal phyla, but is probably best developed in the diatoms and the Chlorophyta. Diatom colonies vary widely in shape, the individual cells in the colony structurally held together to form chains, zigzags, fans, or stars.

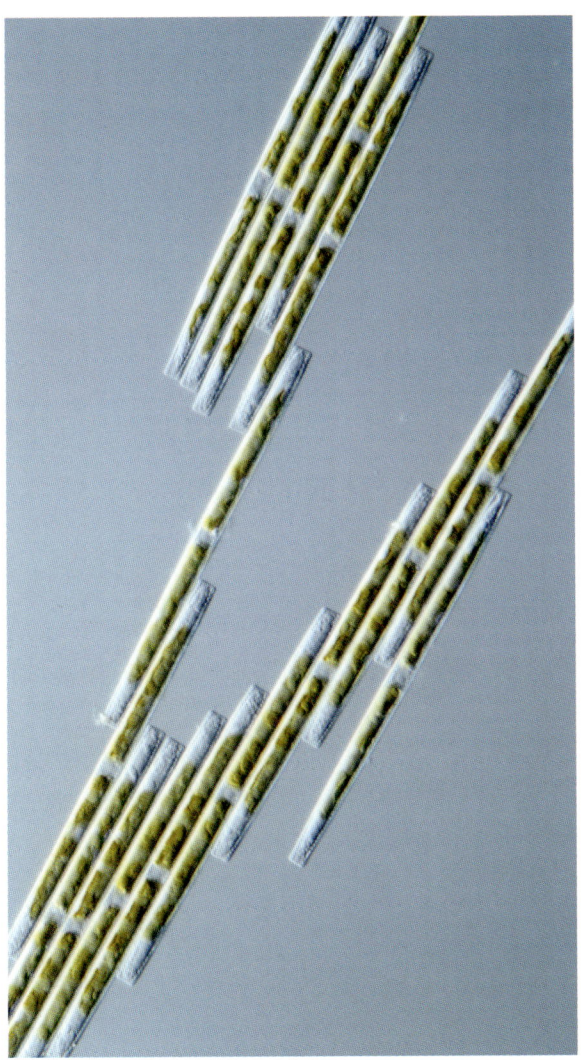

COLONIAL PENNATE DIATOMS

Pennate diatoms also form colonies. Although they lack flagella, some colonies of pennate diatoms are capable of movement. The stacks of 20 or more long thin (linear) cells of *Bacillaria paxillifera* slide back and forth along their length, in a highly distinctive motion. The stiff, stepped chains of *Pseudonitzschia* colonies are formed by the overlapping ends of the adjacent linear cells.

In *Asterionella formosa*—meaning "little star of great beauty"—the linear cells are joined together by mucilage pads at the center of the star-shaped colonies. Attached to seaweed, shells, and stones, cells of *Licmophora* (see page 66) form colonies on branched mucilaginous stalks, the stalks having been secreted by the diatom. The tube-dwelling diatoms *Berkeleya rutilans* and *Navicula ramosissima* each build their habitat, secreting the mucilaginous tubes in which they live. The branched tubes, which often reach lengths of 4 in (10 cm), superficially resemble filamentous macroalgae until the tubes with their resident diatoms are viewed with a microscope.

↖ Long thin cells of the unique colonies of the pennate diatom *Bacillaria paxillifera* slide back and forth along their length to give the colony a distinctive pattern of motion.

← Mucilage pads join the ends of the eight long thin cells of the colonial pennate diatom *Asterionella formosa* into a stellate colony.

COLONIAL CENTRIC DIATOMS

In the centric diatoms, the cells of *Melosira moniloformis* form long chains that are linked together by mucilage pads at the valve center, while a marginal ring of strutted processes connects the cells in *Skeletonema*. In *Chaetoceros* the long projections from adjacent cells interconnect with each other near the cell surface to form the colony. The chains assist with flotation of the nonmotile diatom cells and can also deter predation by zooplankton.

MOTILE COLONIES

The unicellular green alga *Chlamydomonas* is considered the basic unit of the motile colonies of its close relatives, which range from the 16-celled *Gonium* to the 500 to 50,000 cells that construct the supercolonies of *Volvox*. These colonies have long fascinated botanists, who recognized them as important in helping to understand the evolutionary transition from unicellular to multicellular life. For more than 300 years, scientists have marveled at the large green spheres of *Volvox* (see page 103), which are forever revolving forward. It is this behavior that gave the genus its name.

These fascinating microscopic green colonies are composed of *Chlamydomonas* cells, embedded in a mucilage, with the two flagella of each cell protruding from the colony. The number of cells in the colony is genetically determined and fixed early in development. The simple *Gonium* colony consists of a flat layer of 16 cells. However, other colonies are spherical: the solid colony of *Pandorina* has between 8 and 16 cells, while the hollow spherical colonies of *Eudorina* have 16 or 32 cells, *Pleodorina* have 32, 64, or 128 cells, and *Volvox* have 500 to 50,000 cells, all regularly positioned in the extracellular envelope at the colony's periphery. With the exception of *Volvox*, the colony size is limited to 128 cells and 500 microns in diameter—a fascinating conundrum to contemplate.

The *Gonium* colony swims by employing both *Chlamydomonas* and *Volvox* swimming patterns. The beating of the flagella of the four central cells resembles that of *Chlamydomonas*, propelling the cell forward with a slow body rotation. Meanwhile, the flagella of the 12 peripheral cells beat in parallel, similar to that in *Volvox*, which causes a left turn rotation of the colony. In this way, a *Gonium* colony simultaneously swims forward and rotates, whereas a *Volvox* colony only rotates.

→ The colony of the green alga *Gonium pectorale* is a flat plate of 16 cells that are embedded in mucilage and have two whiplike flagella protruding beyond the mucilage.

← Long, thin branched projections (setae) from adjacent cells of the centric diatom *Chaetoceros* aid in flotation by interlocking and joining the cells into a colony.

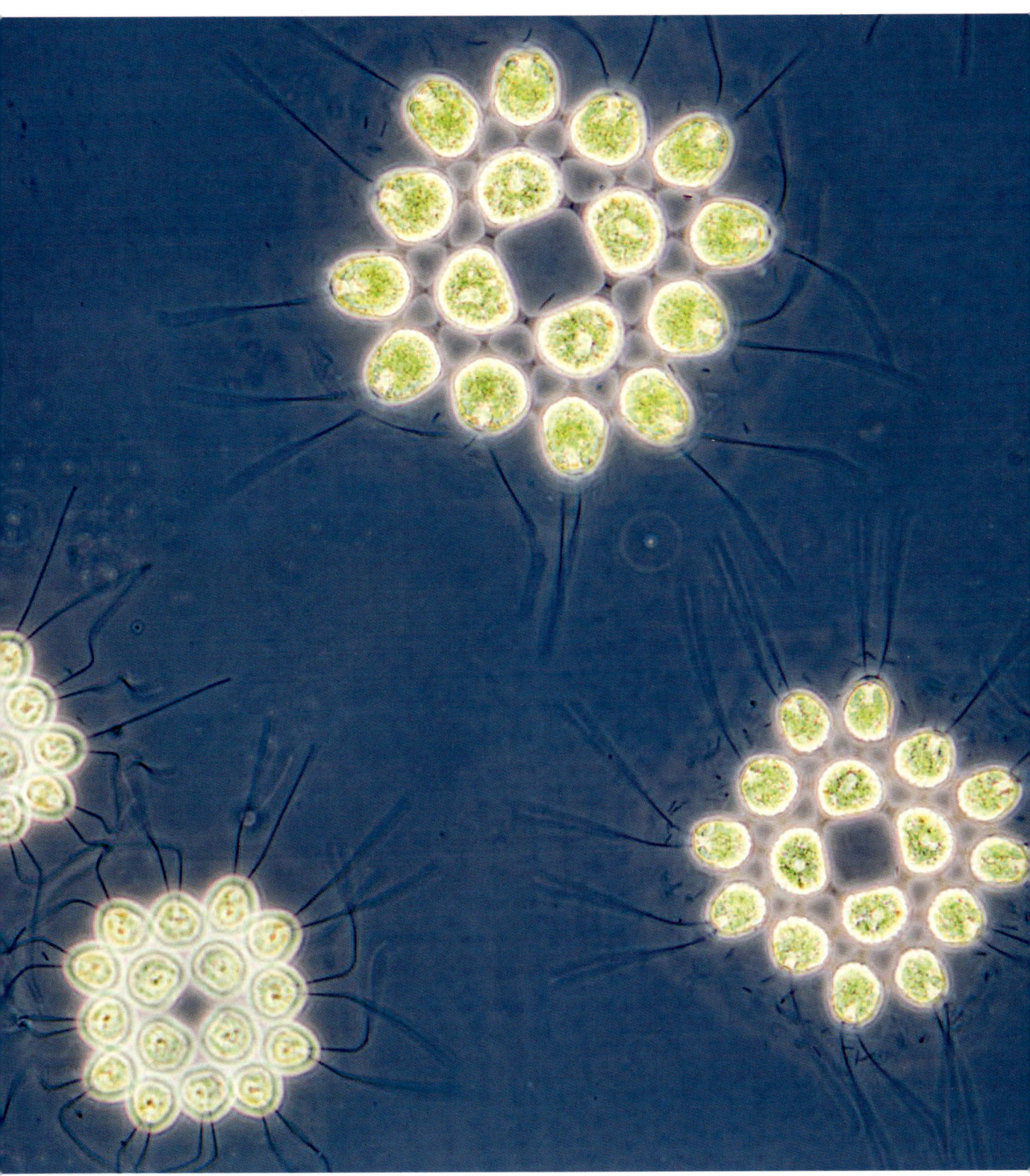

Filaments

A filament is a row of cells in which neighboring cells share a common cell wall. Filaments can be composed of one row, two rows, or multiple rows of cells, and may be branched or unbranched, formed by cell divisions only in one plane (unbranched filaments) or in two planes (branched filaments). The Cyanobacteria, Rhodophyta, Chlorophyta, and Phaeophyceae all have a variety of filamentous thalli.

FILAMENTOUS CYANOBACTERIA

In addition to single cells and colonies, some cyanobacterial species grow as filaments. These can be with or without specialized cells, and are often encapsulated in a discrete extracellular sheath or diffuse slime layer (see page 42). In the genera *Oscillatoria* and *Lyngbya*, the cells in the filament are similar, whereas the filaments of *Dolichospermum* have some specialized cells:

Cyanobacterial filaments
Coiled filaments of *Dolichospermum circinale* (formerly *Anabaena circinalis*) are composed of three cell types: barrel-shaped vegetative cells, spherical heterocysts, and ovoid akinetes.

Heterocyst

Vegetative cell

Akinete

Vegetative cell

akinetes for asexual reproduction and heterocysts for nitrogen fixation. Globally, nitrogen fxation is an important process that converts atmospheric nitrogen gas into ammonia (the nitrogen gas is fixed as ammonia). Many cyanobacterial species contain the bacterial enzyme nitrogenase, which speeds up the rate of the nitrogen fixation reaction. Cyanobacteria use the ammonia in protein synthesis. An anaerobic enzyme, nitrogenase is only functional in the absence of oxygen and, for this reason, is spatially separated in the heterocysts, and protected from destruction by the oxygen released by photosynthesis in the filament's vegetative cells. Some cyanobacterial species that lack heterocysts fix nitrogen only at night when the vegetative cells are not engaging in photosynthesis. Nitrogen fixation is an adaptation that allows cyanobacteria to survive and become abundant in low nitrogen environments.

FILAMENTOUS RED ALGAE

Over 95 percent of the known species of the Rhodophyta—including almost all species of the advanced red algae, the class Florideophyceae—have branched filamentous thalli with either a uniaxial or multiaxial construction, formed by divisions of the apical cells. Species of *Ceramium* are among the many red algal species that have uniaxial thalli. They are often delicate plants composed of a single branched filament arising from one apical cell, with the subapical cells

dividing to form lateral branches. More robust thalli of the red algae are composed of multiple filaments, each filament arising and growing almost exclusively from an apical cell and coalescing to form elaborate multiaxial thalli. In some genera, the filaments become interconnected but remain visible. In still other genera, the filaments are densely compacted and their presence is no longer observable from the exterior of the plant that superficially appears to be composed of tissues.

FILAMENTOUS GREEN ALGAE

Many species of the Chlorophyta have branched and unbranched filamentous thalli. While many species have thalli with uninucleate cells of a normal size, measuring approximately 30 microns long and 20 microns wide, other green algal species deviate from this pattern by having filamentous thalli with large uninucleate cells. Species of the freshwater green alga *Oedogonium* have unbranched filaments in which the large cells—measuring up to 240 microns long and 40 microns wide—are uninucleate and have one large, netlike chloroplast occupying most of the cell face.

↑   Large, colorless cells of a single axial (central) filament are encircled by narrow red bands of small cells; the distinctive bands characterize the red seaweed *Ceramium*.

Siphonocladous thalli

The siphonocladous body plan is best developed in the green algae, in two distantly related groups: the predominantly marine order Cladophorales and the predominantly freshwater stoneworts (phylum Charophyta). The very large multinucleate cells that characterize this body plan have arisen from the repeated divisions of their nuclei without the accompanying cytoplasmic divisions, which usually occurs in many other algal species after nuclear division.

REPEATED NUCLEAR DIVISIONS

In the filamentous green algae *Chaetomorpha* and *Cladophora*, nuclear divisions are not immediately followed by divisions of the cytoplasm as they are in the algae with small uninucleate cells. In fact, nuclear and cytoplasmic divisions in these genera are separate events that result in thalli composed of large multinucleate cells. In some species, such as *Chaetomorpha* (see page 105), the cells are large enough to be visible to the unaided eye. The densely branched filaments of *Cladophora vagabunda* form fluffy or pompom-like thalli, 0.8–2.4 in (2–6 cm) high. The long, cylindrical cells of this species can reach a length of 0.06 in (1.5 mm) and contain more than 100 nuclei as well as numerous small disklike chloroplasts aggregated into a netlike structure around the cell's periphery.

← Visible to the unaided eye, numerous giant cylindrical to club-shaped cells of *Valonia fastigiata* are densely packed into a cushion-like dome.

→ In the *Chara* thallus, the long single-celled multinucleate internodes alternate with narrow multicellular nodes from which arise whorls of lateral branches bearing spherical and ovoid reproductive structures.

CURIOUS CELL DIVISION

In marked contrast to *Cladophora* and *Chaetomorpha*, the immature plants of *Valonia* and *Dictyosphaeria* are composed of a single large multinucleate cell that, as it matures, divides into many smaller cells. They undergo a rather extraordinary mode of vegetative cell division that is unique to *Valonia* and its close relatives. Termed segregative cell division, the single cell spontaneously cleaves at its periphery into many often lens-shaped cells that are retained inside the parent cell.

The *Valonia* plants are not filamentous like those of *Cladophora* and *Chaetomorpha*, but are constructed of numerous, cylindrical to club-shaped cells, which are either tightly or loosely interwoven into cushion-like domes. The giant multinucleate cells of *Valonia fastigiata* can reach a length of 0.8 in (20 mm) and form clumps almost 8 in (20 cm) across.

STONEWORTS

Chara and *Nitella* plants accumulate hard white calcite calcium carbonate on their body surfaces, attracting the colloquial names "stonewort" or "brittlewort." Stoneworts have highly complex thalli. They resemble some aquatic flowering plants that have a similar nodal and internodal organization and whorled branches.

Growing to around 3 ft (1 m) high, the main branches of *Chara* are differentiated into a series of alternating short nodes and long cylindrical internodes. Whorls of lateral branches arise from the nodes. The nodes are composed of small cells, unlike the internodes that have a different cellular organization. The young dividing cells of the internode segments have a single nucleus, but the older cells have a siphonocladous level of organization comprising large multinucleate cells. The internode cells can grow as long as 6 in (15 cm) and contain more than 1,000 nuclei, each replicated from the single original nucleus.

Siphons

The green algal orders Dasycladales and Bryopsidales have a siphonous body plan, the basic unit of which is the siphon—a long tube containing a protoplast of numerous nuclei and chloroplasts. The siphonous body plan has arisen from repeated nuclear divisions, coupled with a complete lack of corresponding cytoplasmic divisions.

SIPHONOUS PLANTS

The siphon is not a cell; there are no cross walls or compartmentalization of the protoplast into cells in the siphonous green algae. Rather, the siphons are acellular—this is life without cells. In species of *Bryopsis* and *Caulerpa* (see page 113), one siphon forms the entire algal body, although a cross wall forms at the base of the reproductive structures in *Bryopsis* and other species of the same family.

Siphonous plants present an alternative scheme to the partitioning of the protoplasm into cells and the development of multicellularity. This partitioning has at least three functions that affect vegetative plants, which have to be dealt with by the siphonous algae.

In multicellular plants, the cell wall and plasma membrane isolate an appropriate volume of cytoplasm around the nucleus. This isolation permits the nucleus to express the genes necessary for cell differentiation. The high plasma membrane surface area to cell volume ratio maximizes the diffusion of nutrients and gases into the cell. The cell walls also distribute the mechanical forces applied to the cell, and maintain the three-dimensional structure.

THE DASYCLADALES

Species of the Dasycladales have a unique body plan. Amazingly, the plant consists of a giant, highly differentiated single cell ranging from 0.04 to 8 in (1–200 mm) in length. Furthermore, the unmistakable radial symmetry of the dasyclad thallus, constructed from the central siphon, and whorls of lateral branches of varying complexity (which are reminiscent of multicellular algae), present a basic body plan from which many more elaborations are possible.

The most famous dasycladalean is *Acetabularia*, whose thallus has been used extensively in research investigating cellular processes. The mature uninucleate plant of *Acetabularia* is composed of a holdfast and a remarkable unbranched central siphon of 4–8 in (10–20 cm) in length, which bears four generations of hairlike whorls of branches and a terminal reproductive cap; this delicate thallus is commonly known as the "mermaid's wine glass." Even more remarkable is the siphon's single giant nucleus, which is 150–180 microns in diameter. This is considerably larger than most green algal nuclei, which typically have a diameter of less than 10 microns.

→ The vegetative thallus of the mermaid's wine glass (the green alga *Acetabularia*) is one giant cell with a single nucleus; its slender central siphon bears successive hairlike whorls of branches (not visible) and a conspicuous terminal reproductive cap.

← The large, much-branched thallus of the spongeweed (*Codium tomentosum*) is made up of microscopic interweaving siphons that are colorless and loosely organized (spongy) in the inner thallus and green, swollen and closely pressed together at the thallus surface.

→ The mermaid's tea cup (the calcareous green alga *Udotea cyathiformis*) is composed of long branched microscopic siphons that are aggregated into a cup, instead of the fanlike thallus more typical of *Udotea* species.

THE BRYOPSIDALES

The siphonous green algae of the order Bryopsidales exhibit an extremely broad morphological diversity, ranging in form from the single siphon in *Bryopsis*, *Derbesia*, and *Caulerpa* (see page 113), through to the elaborate interweaving of branched siphons in the thalli of *Codium*, *Halimeda*, and *Penicillus*.

The single siphon of *Bryopsis* has a simple morphology, comprising a prostrate portion growing horizontally across the rocks, and soft, erect featherlike branched fronds up to 4 in (10 cm) long. By comparison, the much larger and more robust single siphons of *Caulerpa* have a well-developed stolon running across the substratum that gives rise to erect fronds, whose differing form defines the many species.

In some species of the Bryopsidales, the branched siphons may be loosely aggregated—as in the thalli of *Chlorodesmis* and *Boodleopsis*—or tightly aggregated into the fan-shaped thalli of *Udotea* and *Rhipilia*, the globuliferous (pom-pom) thalli of *Tydemania*, and the chainlike segmented plants of *Halimeda*.

The thalli of *Udotea* are composed of an uncalcified basal rhizoidal mass that is responsible for anchoring the plant into the sand, a calcified upright stalk, and a calcified erect fan. In comparison, species of *Halimeda* (see page 146) are composed of a holdfast of disorganized siphons. From these rise linear arrays of flattened calcareous segments, which are variously shaped (often kidney- or heart-shaped) and connected to each other by flexible joints (nodes). Inside each segment there is a well-organized outer region of cortical siphons surrounding the inner medullary siphons, which form the nodes. The tips of the cortical siphons adhere to each other, enclosing the space between the siphons into which aragonite calcium carbonate is deposited. This aragonite provides structural support to the siphons.

Fronds

Fronds are the main part of the seaweed above the holdfast. They are formed by the division of cells in two or more planes to produce undifferentiated plant tissue, generally consisting of spherical or slightly elongated cells. Fronds vary in form from sheets, tubes, sacs, and cushions to large, leathery blades.

SHEETLIKE FRONDS

The simplest fronds are the sheetlike thalli of the primitive red algal genera *Porphyra* (see page 158) and *Pyropia*, and the green algal genera *Ulva* (see page 150) and *Monostroma*. Plants of these four genera vary greatly in shape, from small tufts to long narrow blades, to large, broadly expanded sheets. Thalli can grow to 16 in (40 cm) or more, and are composed of, depending on the species, one or two layers of cells that are identical throughout the thallus.

SACS AND TUBES

Some seaweeds grow as hollow or solid sacs and tubes, attached to rocks by their rhizoids. Hollow sacs are usually filled with seawater, as is the case with the red seaweed, *Halosaccion glandiforme*, which is commonly known as "dead man's fingers." This particular species has a thallus of elongated ovoid sacs up to 6 in (15 cm) long, whereas species of *Botryocladia* (see page 111) often have solid axes bearing multiple hollow membranous vesicles (sacs) that can resemble a bunch of grapes. The single elongated sac of the red seaweed *Gloiosaccion brownii* can grow to just over 6 in (15 cm) and is filled with mucilaginous extracellular material.

← Closely pressed filaments of the dead man's fingers (the red alga *Halosaccion glandiforme*) form water-filled sacs.

Among the brown seaweeds, the thalli of *Colpomenia sinuosa* attain diameters of 1.2–2.8 in (3–7 cm) and grow as bladderlike sacs with a grooved, convoluted surface, reminiscent of a cushion sitting on the rocks. By comparison, *Asperococcus bullosus* has membranous, hollow, irregularly inflated thalli with a length of up to 27 in (70 cm) and a width of 3 in (8 cm), while the thalli of *Scytosiphon lomentaria* (see page 154) are narrow, hollow tubes growing to a length of 12 in (30 cm) and 0.3 in (8 mm) across. In the green algae, some species of *Ulva* grow as hollow, narrow tubes, often reaching 12 in (30 cm) or more. The saccate and tubular thalli are usually filled with seawater, particularly in those species inhabiting the intertidal zone. A reservoir of seawater protects against the higher temperatures and desiccation stress experienced by seaweeds exposed to the harsh atmosphere at low tide.

CRUSTS

Some marine macroalgal species grow as flat, spreading thalli that adhere closely to the underlying rock. These thin crusts are round or irregular in shape, with a maximum thickness of just 0.08 in (2 mm). The crusts of the brown algal *Ralfsia* and the red algal *Hildenbrandia* have a basal layer of cells that is pressed close to the rocks and gives rise to erect coherent filaments. *Hildenbrandia* crusts are typically thin, firm, cartilaginous, and reddish in color, which is easily distinguished from the pink, hard, coralline algae.

Calcareous algae

In some algal species, calcium carbonate in the form of either calcite or aragonite is deposited on the cell walls. Calcification is best developed in the unicellular planktonic Haptophyta (see page 101); in the green algal orders Bryopsidales and Dasycladales; the stoneworts in the Charophyta; and the more than 500 species of coralline red algae.

CORALLINE RED ALGAE

The coralline red algae lay down calcite and magnesium carbonate in their cell walls, which makes the plants hard and stony. As a result of the white calcite surrounding the small cells in the filaments, the coralline thalli take on a pink to purple color, instead of the dark-red color that typifies the red algae.

The coralline algae exhibit two main growth forms, one that is erect, branched, and articulated, and another that is encrusting (see page 109). The articulated coralline algae have crustose bases and erect, branched fronds made up of alternating calcified segments and small noncalcified nodes, the latter forming flexible joints. The encrusting thallus forms a completely calcified crust that is closely appressed to the substratum on which the coralline is growing or, having broken free, the coralline fragment continues growing as a free-living nodule, rolling around the seafloor.

The coralline thallus is composed not of a tissue but of a multitude of precisely organized, narrow-celled filaments pressed closely to their adjacent filaments by the calcite-impregnated cell walls. The cells in these filaments are 5–15 microns in diameter, and are longer than they are wide, causing the filaments to resemble a string of beads or a chain of sausages.

← The thallus of the articulated coralline red algae consists of alternating hard calcified segments and narrow flexible joints; the latter articulate and permit the thallus to bend at the many joints along its length.

→   Lake balls

ALGAL BALLS

It is a well-known phenomenon that certain detached
red, brown, and green macroalgae will continue to
grow as free-living thalli if trapped and submersed
in very sheltered aquatic environments. Rhodoliths
(see page 220), also commonly known by their French
name *maërl*, are the most widely known algal balls.
They are hard and stony and composed of coralline
red algae, which form vast beds around many coasts
worldwide. The rhodolith *Sporolithon durum* can reach
a diameter of 0.6–4.5 in (16–114 mm), with lumpy
and fruticose protuberances on the ball's surface.
Crustose plants of this species that grow on inshore
rock platforms give rise to rhodoliths that are found
rolling on the seafloor to depths of 100 ft (30 m) along
Australia's southern coast. In the northern hemisphere,
the spherical rhodoliths of *Lithophyllum dentatum*
(named for its toothlike protuberances) are found
off the coast of Ireland.

Filamentous algae can also grow into balls.
A vivid "red tide" bloom of the filamentous red
seaweed *Spermothamnion repens* appeared on
a popular surfing beach on Scotland's north coast.
In this instance, the spectacular swathe of small red
balls was probably washed onto the beach from
a nearby sheltered bay.

In one particularly spectacular incident in 1950,
Linda Irvine, of London's Natural History Museum,
reported the stranding of huge numbers of filamentous
Cladophora balls at the high tide mark at Torbay,
England. The balls were found in a band measuring
almost 10 ft (3 m) wide, 9 in (23 cm) deep, and just
under 1 mile (1.6 km) long. Given an average ball
diameter of 1 in (2.5 cm), more than 7 million balls
were estimated to have fouled the beach.

This phenomenon is not confined to coastal
environments. The green alga *Aegagropila linnaei*—
a relative of *Cladophora*—often forms "lake balls"
in freshwater lakes. Dislodged from the substrate, the
water currents and waves in the lake roll and tumble
the algal filaments, generating perfectly formed spheres
of up to 4 in (10 cm) in diameter, with filaments
radiating out from the center.

Kelps

Although there are several orders of large brown algae commonly referred to as "kelps," only species of the order Laminariales are recognized as kelps by phycologists. The sea palm *Postelsia palmaeformis* (see page 216) is a small kelp that grows to less than 3 ft (1 m), whereas the giant kelp *Macrocystis pyrifera* can reach 165 ft (50 m). True kelps are defined by their morphology and life history.

TRUE KELPS

The thalli of the Laminariales are composed of a holdfast, stalklike stipe, and large, flattened, leathery blades. Unique to the Laminariales, the holdfast is composed of numerous tough, branched, fingerlike segments called haptera, which firmly anchor the heavy kelps to the rock substrata. They do this not only mechanically but also chemically. The haptera produce numerous microscopic rhizoids that make physical contact with the underlying rock, filling every crevice and building up an exact impression of the substratum's profile. The rhizoids also secrete large quantities of a strong mucilaginous adhesive, which bonds them to each other and the rock, maximizing the kelp's grip.

Kelps can have a variety of stipes: single, unbranched, much branched, greatly reduced, or none at all. The stipe may also be rigid, holding the blades erect in the water column, or it can be flexible. In the latter case, pneumatocysts—gas-filled floats—are often located laterally on the stipe, or at the base of the blade, raising the fronds into the sunlit upper layers of the sea.

As with stipes, kelp blades vary widely in structure, and include single blades, fan-shaped blades that are dissected longitudinally into segments, and blades with two rows of flattened lateral outgrowths from either side of the central axis.

← The massive holdfast composed of numerous tough cylindrical haptera attaches the huge thallus of the giant kelp (*Macrocystis pyrifera*) to the underlying rock surface.

Kelps grow from the dividing cells located at the junction between the stipe and blade at the thallus base, and from the dividing cells of the outermost layer of the thallus. The kelp thallus is differentiated into three types of tissues: the outer brown layer, a middle layer of colorless cells, and the innermost layer of long, filament-like colorless cells.

The thallus of the giant kelp *Macrocystis pyrifera* (see pages 90–91) is composed of a holdfast, branched stipes, and multiple blades, each with a pneumatocyst at its base. This perennial is capable of living for four to eight years, and over the years the intercalary meristem, located at the juncture of the blade and the stipe, cuts off new stipes and blades, with the blades living for six months. The resulting thallus is a tremendously large, entangled mass of stipes and blades originating from a single blade.

This requires a huge holdfast, which can measure 4–20 in (10–50 cm) high and wide, with multiple haptera, each measuring 0.08–0.16 in (2–4 mm) in diameter.

New *Macrocystis* blades are formed when a single blade that has been cut off from the meristem starts splitting longitudinally, just above the stipe. This eventually cleaves into two new blades, which will divide again in the same manner. When each successive frond splits, the stipes at the blade's base start elongating to accommodate the newly generated blades. This process ultimately results in slender, flexible stipes that can be from 13 to 33 ft (4 to 10 m) long and from 0.1 to 0.3 in (3–8 mm) in diameter. The stipes bear mature leaflike lateral blades that can grow from 12 to 60 in (30–150 cm) in length and from 2 to 6 in (5–15 cm) across. The blade's surface is smooth or roughened with ridges and furrows, and the margins are conspicuously toothed. Elongated ovoid pneumatocysts form between the stipe and the blades, floating the blades across the water's surface.

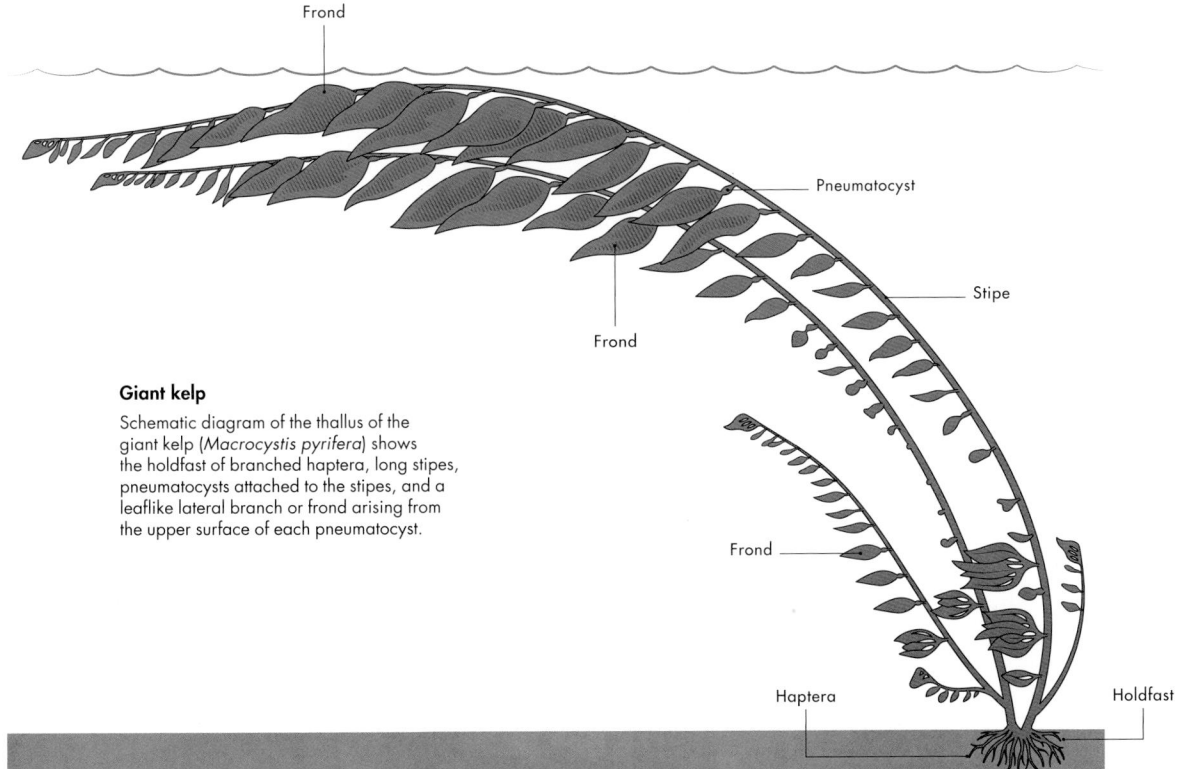

Giant kelp

Schematic diagram of the thallus of the giant kelp (*Macrocystis pyrifera*) shows the holdfast of branched haptera, long stipes, pneumatocysts attached to the stipes, and a leaflike lateral branch or frond arising from the upper surface of each pneumatocyst.

↑ Massive heavy thalli of the giant kelp (*Macrocystis pyrifera*) form submarine forests, the hundreds of fronds on each thallus buoyed upright by a gas-filled pneumatocyst located at their base.

↗ Large gas-filled pneumatocysts, up to 4.7 in (12 cm) long and 1.6 in (4 cm) in diameter, float the upper thallus of the giant kelp at the sea surface where sunlight is at its maximum intensity.

FUCOIDS

The large brown algae of the order Fucales have leathery thalli ranging in size from 4 in (10 cm) to 10 ft (3 m). No fucoid species ever attains the huge dimensions of the giant kelps. Fucoids grow from an apical cell or a group of apical cells sunk in an apical depression, and not from the intercalary meristem, which is characteristic of the kelps. The fucoids and kelps have a diffuse growth pattern, resulting from the divisions of the outermost cell layer of the thallus. Anchored by a discoid holdfast, the fucoid thallus is composed of an outermost brown-colored cell layer, a colorless cell layer, and an innermost colorless cell layer. Many fucoid species, such as knotted wrack (see page 148), have pneumatocysts for flotation.

Species of the order Fucales exhibit considerable morphological diversity. Commonly called wracks, species of *Fucus* (see page 56) have simple, ribbonlike, forked, or flattened thalli, with a midrib. If present, pneumatocysts are embedded into the thallus.

Some fucoids have more differentiated thalli than *Fucus* and the knotted wrack. Thalli of the genera *Cystoseira* and its close relatives and *Sargassum* are composed of a holdfast, and long, erect, cylindrical or flattened primary branches with leaflike laterals that, in some species, have pneumatocysts. A relative of *Cystoseira*, the bladder chain kelp (*Stephanocystis osmundacea*—it's a fucoid, not a kelp as the common name implies) has upper branches composed of up to 12 chainlike, almost spherical pneumatocysts that resemble a string of beads.

→ Robust thalli of the knotted wrack (*Ascophyllum nodosum*) consist of narrow, flattened, leathery primary axes that lack a midrib, branch into two equal forks, and bear prominent egg-shaped pneumatocysts.

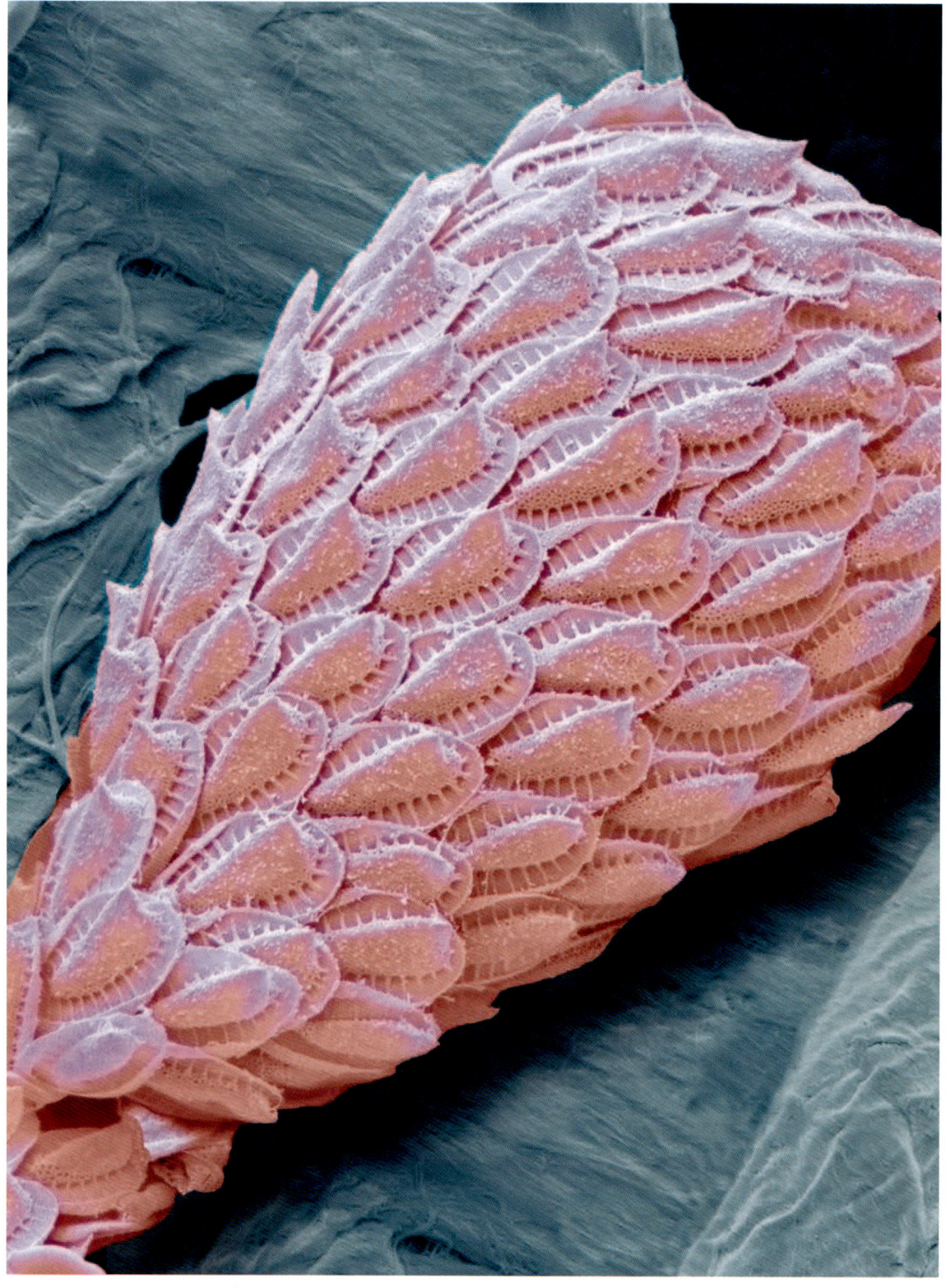

Cell covering

The composition and structure of the cell covering, which varies widely among the algae, is one of the main characteristics used to define different algal phyla (see page 278). Some unicellular algae are naked—defined as having a plasma membrane but lacking a cell wall. However, most algal species form cell walls of varying thickness, which are external to the plasma membrane and often composed of fibers in an amorphous matrix.

CELL WALL COMPOSITION

The most common algal cell wall fiber is the carbohydrate cellulose. This is the main fiber in the thecal plates of the armored dinoflagellates, as well as the cell walls of species of Chlorophyta, Charophyta, Rhodophyta, and Phaeophyceae. In addition to cellulose, the carbohydrates agar and carrageenan are found in the cell walls of red algae, and alginic acid and fucoidan in the cell walls of the brown algae. It is the presence of sulphated carbohydrates, such as fucoidan and agar, that gives many macroalgae a sticky texture.

Other algal phyla have a differing cell wall composition. The frustule of the diatoms is formed of silica (see page 54), as is the cell wall skeleton or scales of some species (*Synura petersenii* and *Mallomonas caudata*; see left and right) of the golden-brown Chrysophyceae. Species of the Cryptophyta and the Euglenophyta have a proteinaceous pellicle for a cell wall, while the Haptophyta have cell walls composed of organic scales or ornate calcareous coccoliths (see page 102). The Cyanobacteria are markedly different to all other algal phyla, as they have a typically bacterial cell wall composed of peptidoglycan, a carbohydrate cross linked by peptides.

← False colored scanning electron microscope image shows the ornate silica scales that completely cover the cell of the unicellular golden-brown alga *Synura petersenii*.

Scales and bristles
Cells of the unicellular golden-brown alga *Mallomonas caudata* are covered by silica scales and long silica bristles.

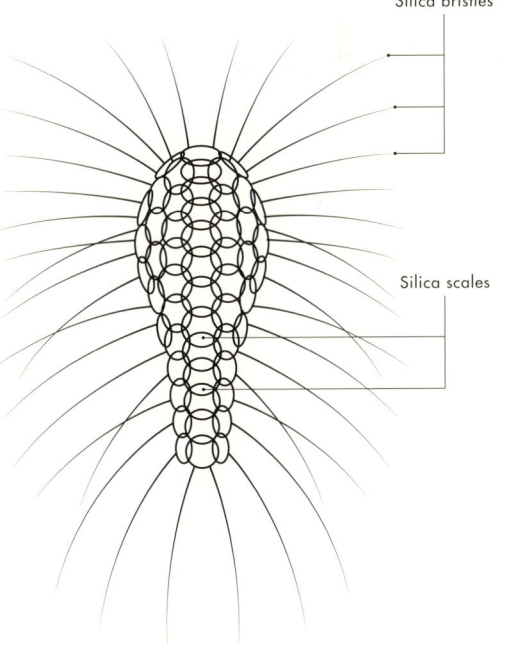

Silica bristles

Silica scales

Motile cells and flagella

In algae, the microscopic motile cells bearing flagella can be vegetative cells or they can be reproductive cells, such as gametes, and zoospores or zoids. While many unicellular algae have flagella, the Cyanobacteria, Rhodophyta, and one class of the charophytes have no motile cells. The brown and green seaweeds only have motile reproductive cells, while in the diatoms, only the male gametes of the centric diatoms have flagella.

FLAGELLAR STRUCTURE

Flagella are whiplike structures (see pages 53, 70, 77) that beat and propel the cell through the water. The eukaryotic flagellum is a complex structure; the part that emerges from the cell is connected to structures inside the cell. The emergent flagellum is made up of microtubules. (These tiny tubes are only visible at the very high magnifications generated in an electron microscope.) They are organized as a central pair of microtubules and a peripheral ring of nine more pairs (the so-called "9+2" complex). The cell becomes motile when the outer nine pairs of microtubules slide past each other, causing the flagella to beat in undulatory or oarlike movements.

The number, form, mode of action, and insertion of the flagella have provided insights into algal evolution and are also used to define the algal phyla (see page 279). Algal cells typically have two flagella that may be inserted into the cell apically (at the apex), subapically (below the apex), or laterally (from the side). The two flagella may be similar in structure (isokont flagella) or dissimilar (heterokont flagella), which is determined by the flagellar length and whether the flagella are smooth or hairy (although flagellar hairs are only visible when cells are viewed in an electron microscope).

Species of the Chlorophyta typically have two similar flagella that are smooth and of equal length, inserted apically into the cell. These cells swim using

"breaststroke"—during the power stroke, the flagella are held nearly straight in front of the cell, and during the return stroke a bending wave begins at the flagellar base and flows to the tip.

The 18 classes of heterokont algae in the kingdom Chromista derived their name from their heterokont flagella: an anterior flagellum bearing two rows of flagellar hairs and a shorter, smooth posterior flagellum. The flagella of phaeophycean motile cells are laterally inserted, while chrysophycean flagella are inserted apically into the cell. In the brown seaweeds, the anterior flagellum generates the propulsive thrust and the posterior flagellum acts as a rudder. The stiff flagellar hairs reverse the propulsive thrust that is generated by the undulations of the anterior flagellum, directing the flow of water to the base of the flagellum and causing the cell to move forward.

In the Cryptophyta, which are common in fresh water, the motile cells have two equal or unequal, apically inserted flagella. They may have one or two rows of flagellar hairs. Unique to the Haptophyta, the haptonema, from which the phylum derives its name, is an appendage and not a flagellum. The haptonema is located between the two nearly equal smooth flagella that are apically inserted into the cell. Varying in length, the longer haptonema detects obstacles and coils; the flagella change their beat, quickly propelling the cell backward. The haptonema also directs food particles down to the cell surface where the particles are ingested.

Enormous diversity of algal flagella

Algal motile cells typically have two flagella, but they differ markedly in their insertion into the cell and whether they have or lack hairs.

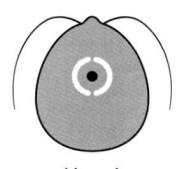

Chlorophyte

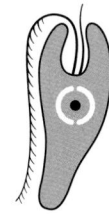

Euglenoid

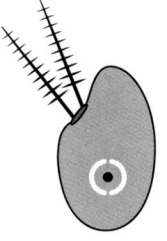

Cryptophyte

Dinoflagellate

The dinoflagellates have two dissimilar flagella. The ribbonlike helical transverse flagellum with a single row of long, fine hairs is located in a groove encircling the cell, and its wavelike undulations propel and rotate the cell forward. The posterior flagellum that emerges from the longitudinal sulcus and trails behind the cell acts as a rudder. In *Euglena,* the flagella are inserted at the base of a flask-shaped opening, called the reservoir, at the anterior end of the cell. There are usually two flagella, both bearing a row of delicate hairs, although one flagellum is usually short and barely extends beyond the reservoir.

The male gamete of *Chara* is unique among the algae, as it resembles the male gamete of the mosses, demonstrating the evolutionary relationship between the charophytes and the land plants. It is narrow, coiled, and streamlined; it has minimal cytoplasm; and it possesses two long and smooth flagella that emerge apically and extend in a posterior direction to encircle the cell.

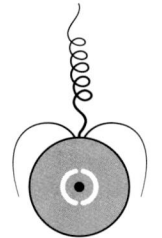

Haptophyte

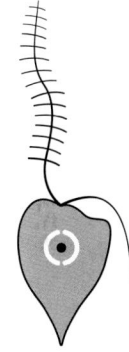

Chrysophyte

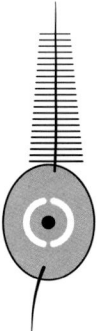

Phaeophyte

Chara

↑ The fan-shaped red seaweed *Martensia elegans* has a pink-blue to purple solid sheetlike lower thallus and a beautifully delicate netlike upper thallus with square to rectangular holes.

↗ The tropical brown seaweed *Padina australis* typically grows as a cluster of fan-shaped fronds that have conspicuous deposits of white calcium carbonate on the upper frond surface.

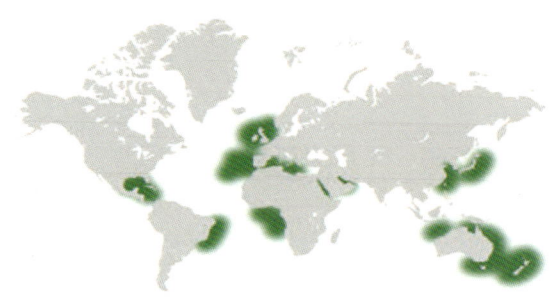

KINGDOM	:	Chromista
PHYLUM	:	Haptophyta
CLASS	:	Coccolithophyceae
ORDER	:	Syracosphaerales
GENUS	:	*Discosphaera*
SIZE	:	Cell plus coccoliths, 15 microns in diameter
HABITAT	:	Planktonic in tropical to warm temperate seas

PHYLUM HAPTOPHYTA

Discosphaera tubifera

Trumpet alga

The spectacular cell covering of *Discosphaera tubifera* is made up of exquisite calcite scales called coccoliths. These scales characterize many species of the golden-brown algal phylum Haptophyta, which are commonly known as the coccolithophorids.

Coccoliths vary widely in structure and are used to define haptophyte species. Platelike coccoliths are produced by many species that organize these calcite scales in differing patterns on the cell surface. Some species assemble far more elaborate coccoliths. *Discosphaera tubifera*, for example, has microscopic trumpet-shaped coccoliths, comprising a basal disk and a tubular stalk supporting a hollow funnel, while *Scyphosphaera apsteinii* has a central cluster of small platelike coccoliths surrounded by a ring of vase-shaped coccoliths, creating a structure that is reminiscent of a flower.

There are two main types of coccolith, both of which form a calcified shell around the cell: holococcoliths form on the cell's surface as numerous tiny calcite crystals, held together by organic material, while heterococcoliths are assembled inside the cell and later secreted at the cell surface as relatively thin, interlocking coccoliths.

The heterococcoliths of *Emiliania huxleyi* (see page 260), a common and widely distributed species, are most likely synthesized one at a time in an intracellular compartment. This compartment is thought to have originated from the Golgi body, a secretory organelle in the cell's cytoplasm consisting of stacked, flat membranous sacs. The process of constructing this species' platelike coccoliths starts with the formation of a coccolith vesicle that lies beside the nucleus. The coccolith vesicle then forms an organic baseplate on which calcite crystals are deposited in an ordered array. The crystals continue growing, forming the coccolith's ring. The completed coccolith is extruded through the plasma membrane and inserted between older coccoliths on the cell's surface.

Calcareous trumpets

Trumpet-shaped coccoliths of *Discosphaerea tubifera* are attached to the cell by the basal disc from which arises a stalk and hollow funnel.

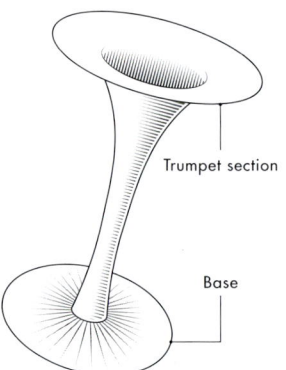

Trumpet section

Base

→ False-colored scanning electron microscope image of the golden-brown alga *Discosphaerea tubifera*, with calcareous coccoliths that completely cover and obscure the small cell that made them.

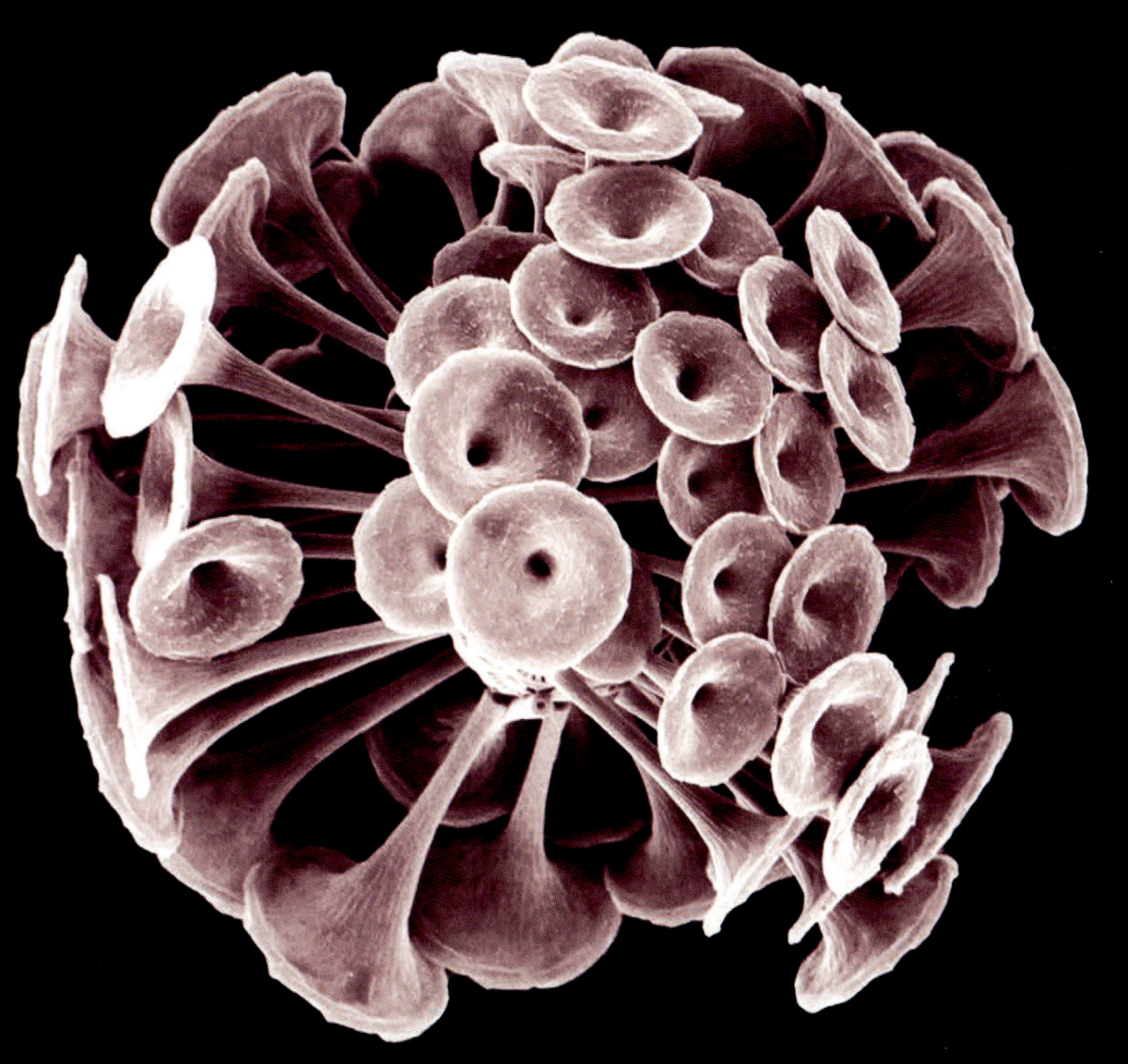

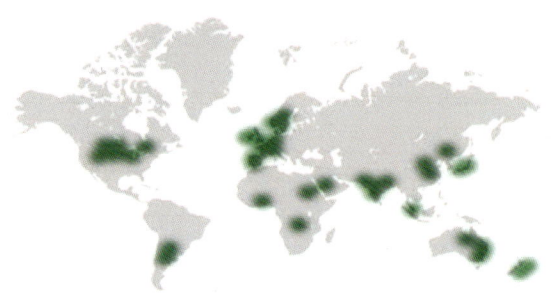

PHYLUM CHLOROPHYTA

Volvox aureus

Volvox

KINGDOM	Plantae
PHYLUM	Chlorophyta
CLASS	Chlorophyceae
ORDER	Chlamydomonadales
GENUS	*Volvox*
SIZE	Colonies 150–3,000 microns in diameter
HABITAT	Planktonic in freshwater lakes, ponds, and ditches

For 300 years, scientists have marveled at the rotating colonies of Volvox. Between 500 and 50,000 biflagellate cells, equidistant to their neighbors and interconnected by fine protoplasmic strands, are positioned radially in the extracellular envelope at the hollow colony's periphery.

The colonies of *Volvox aureus* typically have from 500 to 1,500 cells and measure from 400 to 600 microns in diameter, although some larger colonies are composed of 3,500 cells and are 850 microns in diameter. Each cell bears two flagella that project into the water. The beating of the many cells propels the colony through the water.

Why the supercolonies of *Volvox* evolved is not known, but it could be at least partly related to the colony's motility: a *Chlamydomonas* cell and the smaller volvocalean colonies swim at speeds of 30–275 microns per second, whereas *Volvox* colonies move at 100–600 microns per second. As the Volvocales live in quiet standing waters, the negatively buoyant *Volvox* colonies need motility to stay afloat in the sunlit surface waters. Their motility has the additional advantage of mixing the surrounding water, which increases nutrient delivery to the colony.

As a photosynthetic planktonic organism, the ability of the *Volvox* colony to swim toward the light—a phenomenon known as phototaxis—is central to their survival. However, without a central nervous system, the phototactic steering in *Volvox* is driven solely by the response of the individual cells.

Each cell has a photosensor apparatus (red eyespot) and the tightly associated, light-sensitive pigment chlamyrhodopsin on the plasma membrane above it. The pigment detects when the colony axis is not aligned to the light and signals the flagella to change their beat, rotating the colony toward the light.

Volvox aureus is a widely distributed species that probably has a cosmopolitan geographical distribution pattern. It is a common inhabitant of freshwater lakes, ponds, puddles, and ditches during the late summer in temperate regions.

REPRODUCTION

Volvox colonies produce relatively few reproductive cells, a phenomenon thought to be related to their extraordinarily large colony size. In *Volvox*, 99 percent of cells are incapable of reproduction compared to the smaller colonies (with 128 cells) of its relative, *Pleodorina*, in which only 20–50 percent of cells do not reproduce. *Volvox* colonies reproduce asexually by producing conspicuous spherical daughter colonies and sexually by producing small separate male and female colonies that release spermatozoids and eggs respectively.

→　Numerous cells in the huge motile colonies of *Volvox aureus* appear as small green dots embedded in the peripheral envelope that also encloses several conspicuous daughter colonies inside the hollow parent colonies.

Chaetomorpha coliformis

Green beads

KINGDOM	:	Plantae
PHYLUM	:	Chlorophyta
CLASS	:	Ulvophyceae
ORDER	:	Cladophorales
GENUS	:	*Chaetomorpha*
SIZE	:	Filament to 24 in (60 cm) long
HABITAT	:	Lower intertidal to upper subtidal zones on moderately exposed coastlines

The species *Chaetomorpha coliformis* is a member of the order Cladophorales. This order includes green algal species with branched and unbranched filaments, composed of one row of large multinucleate cells — the siphonocladous level of organization.

Chaetomorpha coliformis usually grows on other plants (epiphyte), such as seagrass and other algae, as a mass of densely clustered filaments that attach to the host plant through a lobed holdfast with a fringed margin. The thallus is an unbranched filament that resembles a string of spherical green translucent glass beads. It is a common species found on the wave-exposed coasts of south eastern Australia and New Zealand, where it grows from low tide mark to depths of 10-13 ft (3-4 m).

The giant cells of *Chaetomorpha coliformis* are the largest in the order Cladophorales, and are exceptional when compared to the cells of many other filamentous algal species, which are invisible to the unaided eye. The multinucleate cells are organized into uniseriate filaments, with the single row of cells increasing in size from 0.02 to 0.04 in (0.5–1 mm) near the filament's base to 0.08 to 0.2 in (2–5 mm) at the apex. These measurements indicate that filament growth happens mainly through intercalary cell divisions close to the filament's base, followed by the enlargement and elongation of the cells. The mid and upper cells of the filament are constricted at their cross walls, giving them their spherical shape, which is maintained by the pressure of the cell contents pushing against the cell wall.

→ Unbranched stiff filaments of the green seaweed *Chaetomorpha coliformis* hang down from their holdfast at low tide, their large cells clearly visible to the unaided eye.

Dividing below

Schematic diagram of the filaments of *Chaetomorpha coliformis* showing the gradual increase in cell size from the filament base to the filament tip.

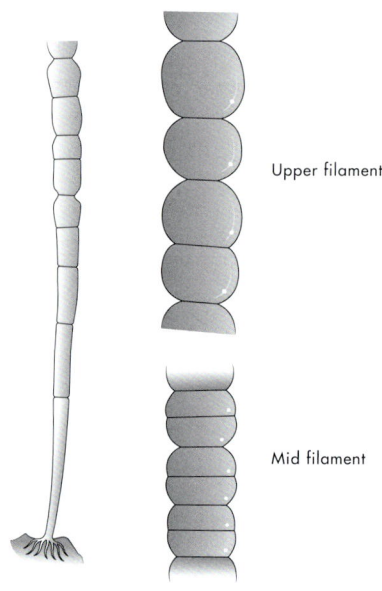

Upper filament

Mid filament

Gelidium elegans

Tengusa

KINGDOM	Plantae
PHYLUM	Rhodophyta
CLASS	Florideophyceae
ORDER	Gelidiales
GENUS	*Gelidium*
SIZE	Fronds 4–12 in (10–30 cm) long
HABITAT	On rocky coasts, subtidal 16–49 ft (5–15 m) deep

The tough, flattened, fernlike fronds of certain species of the genus *Gelidium*, and the closely related genera *Pterocladia* and *Pterocladiella*, are a distinctive thallus form often associated with the order Gelidiales. In these species, the fronds are frequently bipinnately branched.

In pinnate fronds, two rows of lateral branches—or pinnae—arise one on either side of the main axis, a branching pattern similar to that in bird feathers and some ferns. Bipinnate fronds are formed when the lateral branches of a pinnate frond bear small branches or pinnules forming in two rows either side of the lateral branch.

Plants of *Gelidium elegans* (formerly *Gelidium amansii*) are dark red, tough, erect, and densely branched. They can grow to 12 in (30 cm) in length. Both a holdfast and runners (stolons) bearing bundles of rhizoids anchor the thalli to the substratum. The thallus grows by divisions of the apical cell and is composed of filaments that are interconnected to other filaments in a more or less defined manner, giving the plant a uniaxial construction. Defining and identifying species of the Gelidiales based on their highly variable and simple morphology have been notoriously difficult, especially without fertile thalli (which are rare in wild populations). These difficulties are now being resolved by DNA sequencing studies.

FOOD PLANT

The Japanese have long used *Gelidium elegans*, commonly known as tengusa, as a food plant. They extracted the carbohydrate agar from the plants' cell walls and made the product kanten, which has gelling properties similar to gelatine. Since as early as 750 CE, the Ama, freediving fisherwomen who can hold their breath for two minutes, have harvested the tengusa from its natural habitat, the rocky seafloor at depths of 16–50 ft (5–15 m). They wore only a loin cloth, a bandana, and a rope around their waist securing them to the boat. Goggles were introduced in the early 1900s and wet suits and flippers followed but, in order to conserve this natural resource, scuba tanks have never been used. The tradition of the Ama survives to the present day.

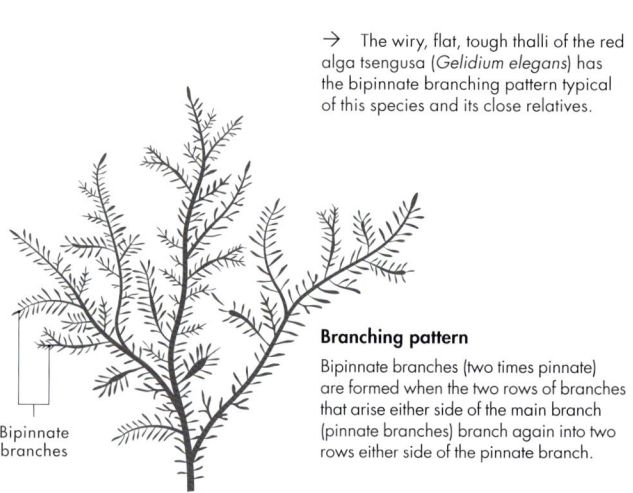

→ The wiry, flat, tough thalli of the red alga tsengusa (*Gelidium elegans*) has the bipinnate branching pattern typical of this species and its close relatives.

Bipinnate branches

Branching pattern
Bipinnate branches (two times pinnate) are formed when the two rows of branches that arise either side of the main branch (pinnate branches) branch again into two rows either side of the pinnate branch.

KINGDOM	Plantae
PHYLUM	Rhodophyta
CLASS	Florideophyceae
ORDER	Corallinales
GENUS	*Crusticorallina*
SIZE	Crusts up to 1.75 in (4.5 cm) in diameter
HABITAT	Epilithic, from the mid-intertidal zone to a depth of 45 ft (14 m) on exposed and protected shores

PHYLUM RHODOPHYTA

Crusticorallina muricata

Coralline red alga

Plants of the encrusting coralline red alga *Crusticorallina muricata* form loosely attached, undulating crusts that overgrow the contours of the underlying rock, other encrusting algae, and kelp holdfasts. The pink crust has a smooth surface with white swirls, white margins, and is relatively thick (350–950 microns). Typical of the advanced red algae, the crusts are composed of tightly packed filaments.

Crusts are one of the two morphological forms recognized for the coralline red algae. The other, the articulated form, has crustose bases and alternating calcified and noncalcified erect segments. These two different morphological forms are classified in different orders of the red algae. It has long been established, and supported recently by DNA studies, that evolution has progressed from the crusts to the articulated species.

First described in the early 1900s, the species currently known as *Crusticorallina muricata* has been studied several times in the intervening period. In 1970, an eminent coralline taxonomist considered that the anatomy of the species was peculiar. However, no phycologist suspected what recent DNA sequencing studies subsequently revealed. The one species recognized for a hundred years was actually four species of the new genus *Crusticorallina*. Even more astounding, the four *Crusticorallina* species were genetically related, not to the other crustose species, but to the articulated species, necessitating their classification as the only crustose genus in a subfamily of articulated coralline red algae. The DNA data indicated that *Crusticorallina* had evolved from an articulated coralline ancestor, representing an evolutionary reversal, the loss of the erect articulated part of the thallus and a return to only the crust.

Crust anatomy

Schematic diagram through the margin of an encrusting coralline red alga (*Lithophyllum*) showing the filamentous construction and growth from the meristems.

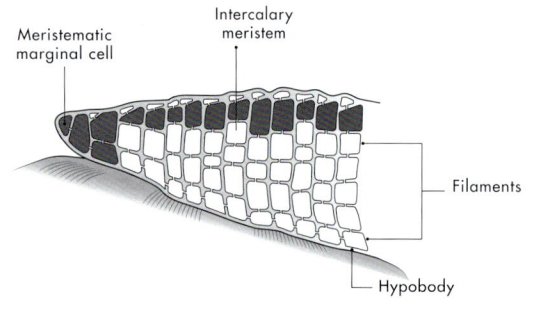

Meristematic marginal cell

Intercalary meristem

Filaments

Hypobody

→ Limey pink crusts of *Crusticorallina muricata* with their conspicuous white margins are an evolutionary reversal, having evolved from an articulated coralline ancestor.

Botryocladia leptopoda

Grape weed

KINGDOM	Plantae
PHYLUM	Rhodophyta
CLASS	Florideophyceae
ORDER	Rhodymeniales
GENUS	*Botryocladia*
SIZE	Thalli 4–12 in (10–30 cm) tall
HABITAT	Grows on rough-water coasts, from low tide mark to 80 ft (24 m)

Species of *Botryocladia*, such as the commonly known grape weed, include some of the most beautiful and easily recognizable red seaweeds. Many have an elegant and distinctive morphology comprising erect solid cylindrical axes covered with vesicle-like lateral branchlets.

The vesicles of grape weed are hollow, swollen, and filled with mucilage. Although superficially appearing to be composed of tissues, the plants of *Botryocladia* are actually filaments that have combined to form an elaborate multiaxial thallus. The bright red plants of grape weed are 4–16 in (10–40 cm) tall, held erect by the solid, rigid, cartilaginous axes. The numerous hollow, ovoid vesicles, ranging from 0.2 to 0.5 in (5–15 mm) long and from 0.12 to 0.24 in (3–6 mm) in diameter attach by a short stipe to the axes, almost covering them completely to somewhat resemble a bunch of grapes.

Species with grapelike branches and vesicles conform to the generic name derived from the Greek *botrys* (a bunch of grapes) and *clados* (a branch), and to the common names sea grapes or grape weed. Various *Botryocladia* species from the Mediterranean Sea, the eastern Pacific coasts, and from the tropical western Atlantic also have long robust axes

clothed in masses of vesicles. However, not all species of *Botryocladia* have long axes. Some species bear either small near-spherical or large elongated cylindrical vesicles on inconspicuous short axes.

Approximately 40 species of *Botryocladia* have been described worldwide. Many of these are widely distributed in tropical to warm temperate seas, where they typically inhabit the deep shade under ledges in lower intertidal rock pools and more commonly in shady habitats below low tide mark and in the deep subtidal zone. One species, *Botryocladia botryoides*, lives in caves and shady crevices in the intertidal zone and in the subtidal zone to 56 ft (17 m) deep.

→   The solid stemlike axes of the grape weed, the red seaweed *Botryocladia leptopoda*, are densely covered with vesicles.

Caulerpa sertularioides

Feather Caulerpa

KINGDOM	Plantae
PHYLUM	Chlorophyta
CLASS	Ulvophyceae
ORDER	Bryopsidales
GENUS	*Caulerpa*
SIZE	Fronds 1–3.5 in (2.5–9 cm) long
HABITAT	Rocky, sandy, muddy tropical shores, intertidal to depths of 80 ft (25 m)

The plant body of *Caulerpa* is composed of a single siphon that can reach almost 10 ft (3 m) in length in some species. Within this remarkable and unique siphon, the multinucleate protoplast—with its many chloroplasts and other organelles—functions as a single entity. There are no cells.

Although one long tube, the *Caulerpa* thallus has differentiated into three structures: a stolon, rhizoids, and erect fronds. The creeping, stemlike stolon periodically produces bundles of colorless rhizoids that anchor the plant to the substratum. The stolon also gives rise to erect photosynthetic fronds that vary markedly in size and form, providing characters useful for defining the various *Caulerpa* species. The fronds may resemble feathers (*Caulerpa sertularioides*), or cacti (*Caulerpa cactoides*), or the leaves of a cypress pine tree (*Caulerpa cupressoides*), or they may have a distinctive morphological feature such as serrated margins (*Caulerpa serrulata*), or five to eight longitudinal rows of small, spherical branchlets (*Caulerpa lentillifera*).

Trabeculae are a feature that is unique to *Caulerpa*. They are branched cylindrical ingrowths of the siphon wall that traverse the central cavity of the siphon. These growths extend from the siphon wall deep into the cytoplasm, and are thought to be an adaptation to overcome the lack of internal partitioning provided by the cell walls of multicellular algae. By providing mechanical support, the trabeculae strengthen the siphon and maintain the shape of the erect fronds.

As the trabeculae are enclosed by the plasma membrane, they may also act as diffusion channels for the transport of nutrients and gases from the environment deep into the cytoplasm of the siphon.

Feather *Caulerpa* typically grows as small to large clumps although it sometimes forms extensive mats. The smooth stolons measure 0.02–0.12 in (0.5–3 mm) in diameter and, together with the rhizoids, they anchor the plants to the sandy and muddy substrata. The featherlike fronds vary in color from light green to deep green, and range in size from 1 to 3.5 in (2.5–9 cm) long and from 0.24 to 0.63 in (6–16 mm) across.

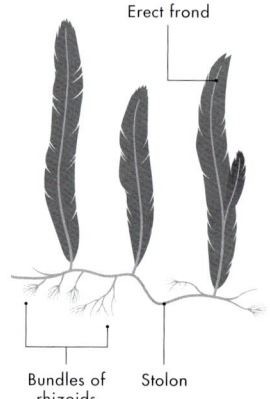

Erect frond

Bundles of rhizoids

Stolon

Creeping alga
The horizontal stemlike stolon of *Caulerpa* species creeps across the seafloor, periodically giving rise to bundles of rhizoids and erect fronds.

→ The dense mat of *Caulerpa sertularioides* is formed by numerous featherlike erect fronds with very fine cylindrical branchlets.

CLASS PHAEOPHYCEAE

Nereocystis luetkeana

Bull kelp

KINGDOM	Chromista
PHYLUM	Ochrophyta
CLASS	Phaeophyceae
ORDER	Laminariales
GENUS	*Nereocystis*
SIZE	Thallus to 115 ft (35 m) long
HABITAT	Subtidal rocky coasts to depths of 100–115 ft (30–35 m)

***Nereocystis luetkeana,* known commonly as bull kelp, is an annual species that appears in subtidal rocky habitats in the spring (March–April), begins reproducing in summer, and deteriorates after ending reproduction in the fall (September–October).**

In the early spring, the young thalli of bull kelp have developed a simple holdfast, stipe, and blade. Two weeks later, the pneumatocyst starts to form from gas released into an internal tear along the transition zone between the stipe and the blade. The pneumatocyst and the blades grow larger, and the stipe elongates upward toward the water surface at a rate of almost 6 in (14 cm) per day.

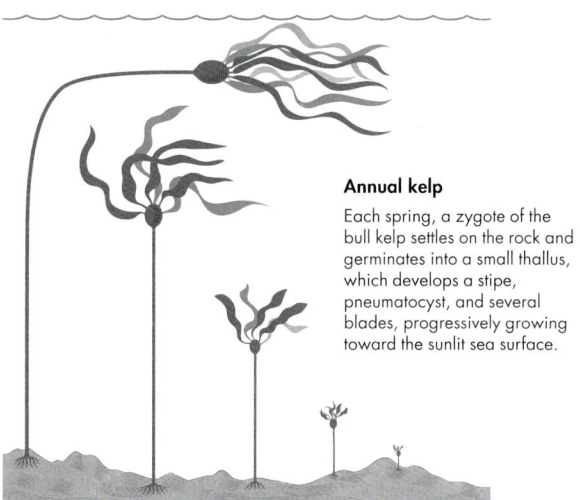

Annual kelp

Each spring, a zygote of the bull kelp settles on the rock and germinates into a small thallus, which develops a stipe, pneumatocyst, and several blades, progressively growing toward the sunlit sea surface.

Each mature thallus is anchored by a holdfast of tough branched haptera from which arises a single flexible, solid, cylindrical stipe that can grow to 115 ft (35 m) in length and approximately 0.8 in (2 cm) in diameter. The stipe increases to a diameter of 2.75 in (7 cm) and becomes hollow a short distance below a single large spherical or ovoid pneumatocyst. This hollow lumen can grow up to 5.5 in (14 cm) in diameter and is filled with a gas that can be 12 percent carbon monoxide.

The 180 cubic inches (3,000 cubic cm) of gas in the hollow section of the stipe and the pneumatocyst provides buoyancy for more than 50 flat, ribbonlike blades that are attached by stalks to the top of the pneumatocyst. The blades, which attain lengths of 13 ft (4 m), are positioned by the pneumatocyst to float parallel to the water surface and spread out just below it, to maximize the capture of light for photosynthesis.

Creating a gas-filled pneumatocyst that ascends through the water column for more than 100 ft (30 m) is a truly remarkable feat, especially when it is potentially experiencing a change in hydrostatic pressure from 4 to 1 atmosphere. However, the *Nereocystis* pneumatocyst successfully maintains an internal pressure of less than 1 atmosphere during its ascent, which is enough to prevent it from buckling.

→ The single gas-filled pneumatocyst of the bull kelp suspends the long flexible stipe upright in the water column and, to maximize photosynthesis, spreads the kelp's long blades out below the sea surface.

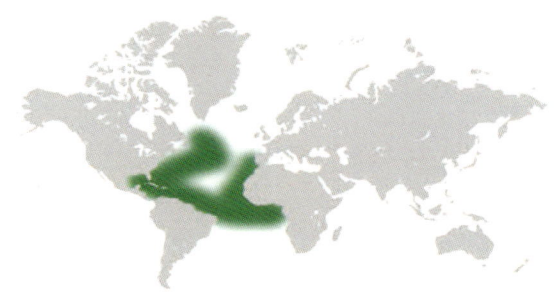

CLASS PHAEOPHYCEAE

Sargassum fluitans

Leathery seaweed

KINGDOM	Chromista
PHYLUM	Ochrophyta
CLASS	Phaeophyceae
ORDER	Fucales
GENUS	*Sargassum*
SIZE	Thallus grows to 3.3 ft (1 m) long
HABITAT	Coastal regions and open seas in some areas of the Atlantic Ocean

Species of *Sargassum* have the most highly differentiated thalli among the algae. Measuring from as little as 4 in (10 cm) to 6.5 ft (2 m) in length, these elaborately branched, leathery, brown algae are differentiated into a holdfast, stemlike stipes, long primary branches, leaflike lateral branches, pneumatocysts for flotation, and reproductive structures (receptacles and conceptacles).

Species of *Sargassum* exhibit distinct seasonal growth patterns and are "pseudoperennials," retaining their slow-growing holdfast and several short stipes but forming new primary branches each year. At the start of the growing season, new cylindrical or compressed primary branches grow from the apex of each stipe. These primary branches, which may be simple or branched, have relatively narrow or leaflike lateral branches, often with a midrib and smooth or serrated margins. Spherical pneumatocysts usually occur close to or in the axils between the primary branch and the lateral branches. Fertile thalli have clusters of receptacles—each typically measuring less than 0.4 in (1 cm) long and 500 microns wide—that form in the axils of the lateral branches. At the end of each season's reproduction, the primary axes are lost from the thallus, leaving only a scar on the stipe.

The light- to golden-brown leathery thallus of *Sargassum fluitans* consists of smooth, cylindrical primary branches up to 3.3 ft (1 m) long. These bear multiple narrow, lance-shaped lateral branches along their length, which can grow up to 2.36 in (6 cm) in length and 0.31 in (8 mm) across. Ovoid to spherical pneumatocysts with a diameter of 0.16–0.2 inches (4–5 mm) are also clustered along the primary branches. This leathery seaweed is a pelagic (lives in open seas) species, forming vast floating meadows in the Atlantic Ocean. Owing to its pelagic habitat, its thallus lacks a holdfast and sexual reproductive organs.

→ The leathery thallus of *Sargassum* is densely branched; each primary branch is crowded with leaflike lateral branches and many small spherical pneumatocysts.

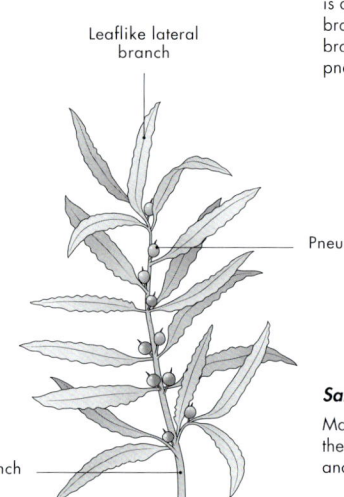

Leaflike lateral branch

Pneumatocyst

Primary branch

Sargassum thallus
Many small pneumatocysts form in the axils between the primary branches and the leaflike lateral branches.

LIFE HISTORY

Reproduction

Some algal species have never reproduced sexually, while others have lost this capacity. However, the majority engage in an intriguing mix of asexual and sexual reproduction, including species whose thalli die in the process of having sex.

LIFE HISTORIES

A life history includes all of the developmental and reproductive events that occur during the lifespan of a species. These events vary widely among the many lineages in the megadiverse algae and are important criteria that are used to define the various algal phyla, orders, genera, and species. Life histories are vital for the long-term survival of a species, as they generate new individuals and resistant stages that ensure the species can overcome adverse environmental conditions. The reproductive cells, gametes (see page 124), and the many different types of spores are the dispersal mechanism in the life history, crucial for a species to colonize new habitats.

HEDGING BETS

The majority of algal species hedge their evolutionary bets by combining various modes of asexual and sexual reproduction. Asexual reproduction produces progeny that are identical to the sole parent, which contributes little to increasing genetic variation in the population, whereas sex has the adaptive advantage of genetic recombination and repair.

The majority of algal species reproduce sexually and usually invest considerable energy in the process. This is apparent during the fertile season of the brown seaweed *Fucus vesiculosus,* in which the tips of the thallus branches enlarge to form conspicuous and swollen reproductive structures. There are, however,

some algal species that have never reproduced sexually. The most notable exceptions are the Cyanobacteria that evolved before the evolution of sexual reproduction. Only eukaryotes reproduce sexually. There are also some unicellular algal species that are reported to reproduce solely by mitosis, as well as some macroalgal species that appear to have lost the ability to reproduce sexually. However, establishing the absence of sex in a species is far from straightforward, as sexual reproduction in algae is often a cryptic process: it can be difficult to recognize, it can be briefly seasonal, and it can feature gametes of the opposite mating types that are not only hard to distinguish from each other, but also from asexual spores.

Curiously, geography may also play a part in controlling the mode of reproduction. Distinct geographical variations in reproductive mode occur in some species that reproduce sexually throughout most of their geographical range except near the range boundaries, where only asexually reproducing populations of the species are known.

→ Swollen fertile areas (receptacles) of the bladder wrack (*Fucus vesiculosus*) develop at the thallus tips. Their mottled appearance indicates the presence of mature reproductive structures in sunken flask-shaped pits (conceptacles) inside the receptacles.

Asexual reproduction

Asexual reproduction, which increases the number of progeny without involving any cellular or nuclear union, is widespread in the algae. There are numerous and varied modes of asexual reproduction employed by different species, including cell division in unicellular species, and vegetative reproduction and asexual spores in multicellular species.

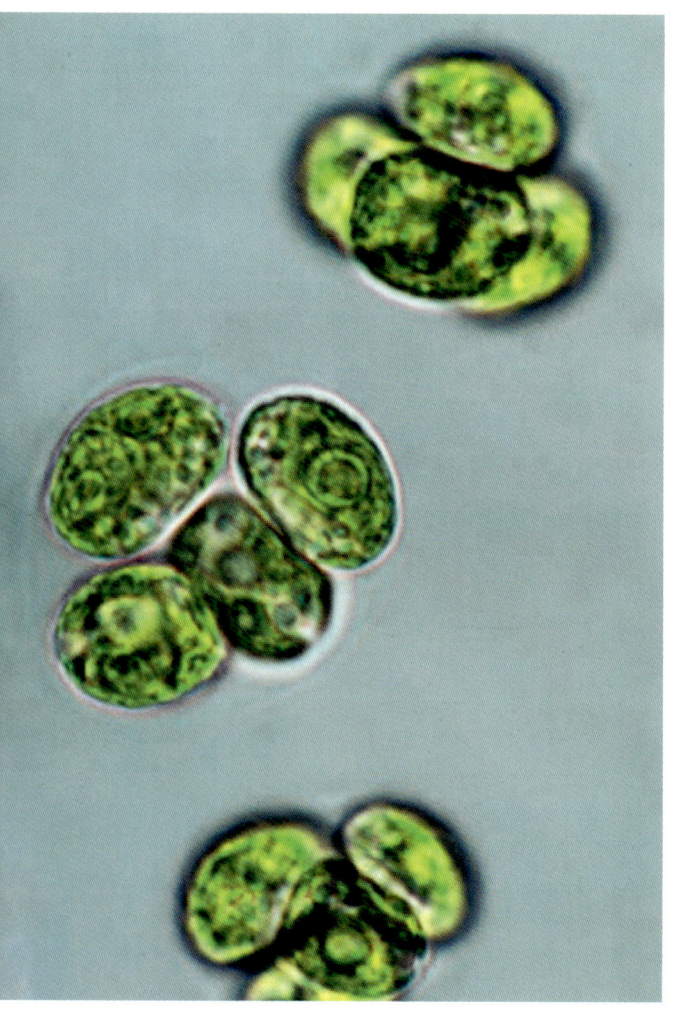

CELL DIVISION

Lacking sexual reproduction, the cells of the Cyanobacteria divide by binary fission, while the unicellular green alga *Chlorella* produces four or eight cells through two or three successive mitotic divisions. In contrast, many unicellular algae reproduce both asexually and sexually, including some species of the green alga *Chlamydomonas*. Vegetative cells of *Chlamydomonas* produce from two to sixteen cells through one to four successive mitotic divisions. When the division rounds have been completed, the parental cell wall breakdowns and the progeny are released into the water.

← Newly divided packets of four microscopic cells of *Chlorella vulgaris* are released as single cells into the surrounding freshwater following the rupture of the transparent parental cell wall.

↗ Vegetatively reproducing thalli of the green seaweed *Halimeda incrassata* (background and center) produce underground rhizoids that grow through the sand and have given rise to the two small plants in the foreground.

VEGETATIVE REPRODUCTION

Vegetative reproduction contributes significantly to maintaining the population size of macroalgal species, producing new individuals from regenerating thallus and cell fragments, outgrowths of the rhizoids, or from special propagules (purposely produced vegetative structures). Macroalgae frequently lose thallus fragments during violent storms and herbivore grazing, and use this fragmentation to varying degrees as a method of asexual reproduction. In the red filamentous alga *Rhodochorton purpureum*, fragmentation is the principal mode of reproduction and is more important than the species' infrequent sexual episodes. Fragments generated by intense herbivore grazing constantly mowing down this turf alga have a high regenerative capacity, and quickly form both adhesive rhizoids that anchor the fragment to the substratum and new filamentous shoots.

The siphonous green macroalga *Bryopsis plumosa* reproduces vegetatively when damaged. It oozes spheres of protoplast that regenerate into new thalli.

The green macroalga *Halimeda* reproduces vegetatively by fragmentation as well as by elongating rhizoids that run through the sand and intermittently produce new plants away from the parent plant.

Macroalgae also reproduce asexually with special spores and propagules. Among the cyanobacterial species are some that produce resistant akinetes (see page 78), which are essentially vegetative cells with a thick cell wall. The macroalgae produce a variety of asexual spores that lack flagella in the red algae or are flagellate in the green and brown algae. Uniquely, one group of filamentous brown algae (in the order Sphacelariales) form triangular propagules 5 to 10 cells long that fall from the parent thallus onto the seafloor and grow into new thalli. Various algal species use these different modes of asexual reproduction to maintain or increase their population size without having to incur the risk or cost of sexual reproduction.

Sexual reproduction

Two key events—meiosis and fertilization—define sexual reproduction. Meiosis or reduction division, which halves the chromosome number to the haploid state (n), alternates with fertilization, which restores the chromosome number to the diploid state (2n). Although these two events occur in all sexually reproducing algae, the structure of the reproductive organs and their gametes and spores vary enormously among the algal phyla.

THE GAMETES

Algal gametes differ greatly in morphology among the various algal phyla. There may also be differences between the male and female gametes of a species. Sexual reproduction may involve isogamy, anisogamy, or oogamy. In isogamy, the male and female gametes are the same shape and size, which often makes distinguishing between the two gamete types difficult, if not impossible, whereas in anisogamy, the female gamete is larger than the male gamete. In oogamy,

the female gametes are nonmotile eggs and the male gametes are flagellate cells (spermatozoids). Evolution proceeded from the primitive isogamy, through anisogamy, to oogamy, with this sequence evolving independently in the various algal phyla.

GREEN SEAWEED SEX

Gametes are formed in gametangia, which are reproductive structures borne on thalli known as gametophytes. Unlike the brown and red seaweeds, no specialized gametangia are formed by species of the green seaweeds. Rather, the vegetative cells of the gametophytes simply divide to form the gametes. The siphonous green algae, which lack cells, convert part or the whole protoplast of the siphon into gametes. In *Bryopsis*, cross walls (septa) partition the lateral branches at their base from the rest of the siphon, transforming each branch into a gametangium.

Isogamy

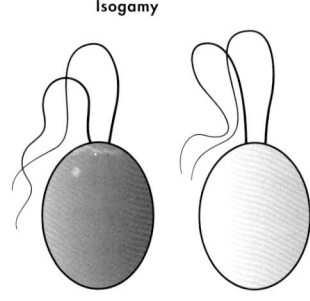

Anisogamy

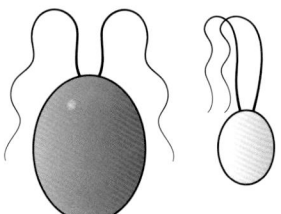

Oogamy

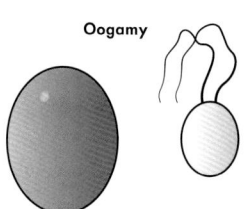

Sexual reproduction and gamete morphology

Within an algal species, the gametes may be equal in size and shape (isogamy), the female gamete (shaded) may be larger than the male gamete (anisogamy), or the female gamete may be an egg and the male gamete a spermatozoid (oogamy).

↑ During sexual episodes in *Halimeda*, the green thallus segments turn white following the transfer of the protoplast into newly formed gametangia—the dark-green outgrowths of the siphons.

In *Halimeda*, the gametangia lack septa and are outgrowths of the siphons located on the edge of each segment (see page 147). The green seaweeds produce gametes with two flagella that are identical in morphology and size (isogametes).

In addition to gametophytes, some green algal species have sporophytes—thalli of the spore-bearing second phase in the life history. The thallus cells of the sporophyte divide to produce zoospores, which have four flagella. It is possible to distinguish between the gametophytes and sporophytes of a green algal species by their biflagellate gametes and quadriflagellate zoospores.

Sexual reproduction in *Bryopsis*

Short lateral branches of the featherlike thallus are transformed into gametangia when a cross wall (septum) forms at each branch base and separates it from the rest of the siphon. Gametes exit the gametangium through a pore.

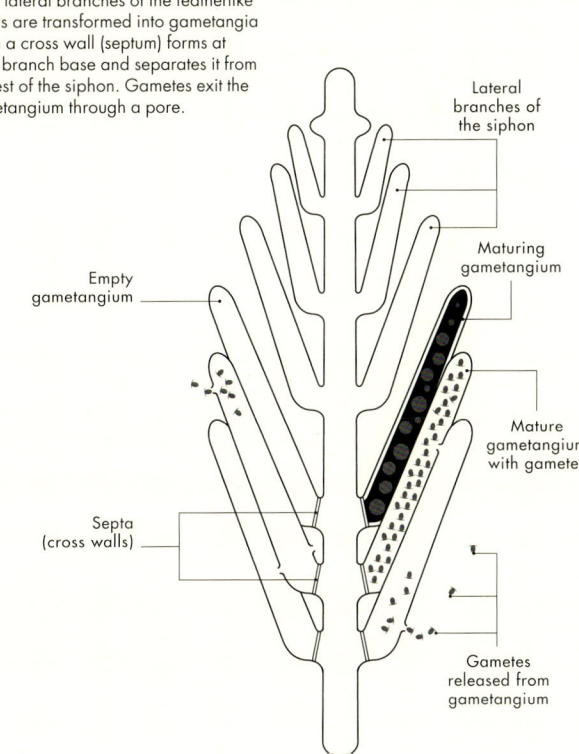

Lateral branches of the siphon

Maturing gametangium

Mature gametangium with gametes

Gametes released from gametangium

Septa (cross walls)

Empty gametangium

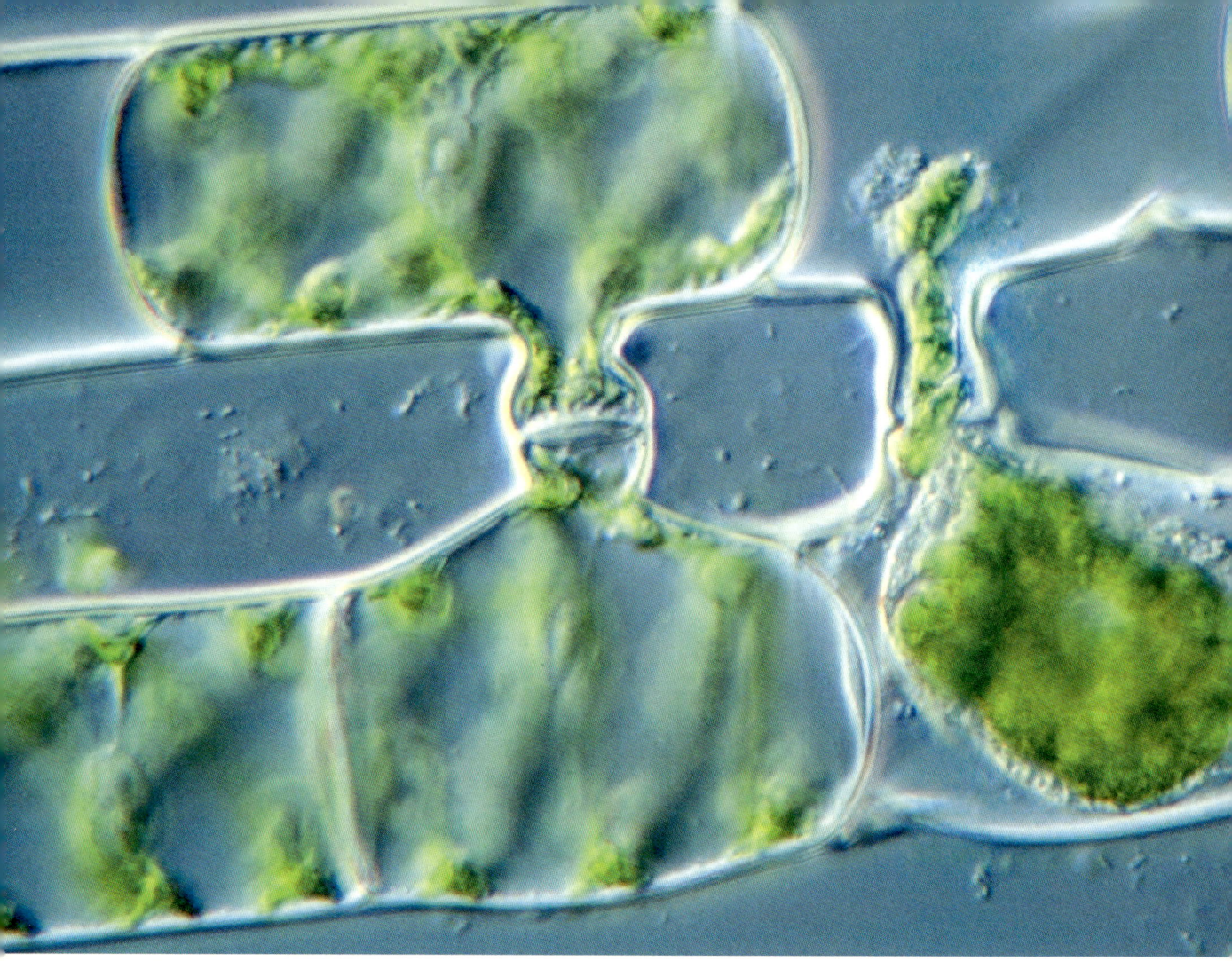

CONJUGATING ALGAE

Two groups of green algae within the Charophyta derive their common name—the conjugating green algae—from their unique mode of sexual reproduction: conjugation. The filamentous species of *Zygnema* and *Spirogyra*, and the unicellular desmids, form no specialized gametangia or gametes. Instead, sexual reproduction is achieved by the fusion of two protoplasts, which is executed via a conjugation tube. Conjugation starts with the filaments coming into close contact, or with aggregations of cells of the desmids. A pair of cells of the opposite mating types align themselves closely to one other, and a conjugation tube grows out from each cell. The ends of these tubes fuse together to form an open tube and

the protoplasts (essentially the gametes) of the two mating types can then either move into the conjugation tube and fuse, or one protoplast can move through the tube to form a zygote in the second cell.

BROWN SEAWEED SEX

In contrast to the green macroalgae, the brown seaweeds form differentiated reproductive structures. Two types are recognized: plurilocular gametangia characterize the primitive brown seaweeds, and unilocular sporangia. Plurilocular gametangia are microscopic and conical to spherical structures attached singly or in loose groups to the main thallus axes. These reproductive structures are divided into many

locules (compartments) that are organized into tiers, each locule containing one gamete. In the oogamous brown algae, the female gametangium or oogonium typically forms between one and eight spherical ova in each oogonium, while the male gametangium or antheridium releases up to 64 spermatozoids—the precise number of each depends on the species. Similar to the Chlorophyta, some brown algal species also have a sporophyte phase in their life history. These sporophytes bear unilocular sporangia that usually release 64–128 motile reproductive cells called zoids.

GAMETE RELEASE

In the algae, gametes are discharged from mature gametangia all at once. The explosive release is often triggered by an external stimulus such as light, mechanical stress, or a sexual pheromone; if light is the stimulus, gamete release usually happens within 30 minutes after dawn. Gamete release often occurs in one direction, with the gametes exiting through a newly formed pore more quickly than the gamete can swim on its own.

↑   The filamentous green alga *Spirogyra* reproduces sexually when two filaments of the opposite mating types pair and produce a tube through which the protoplast of one cell moves to fertilize the other cell.

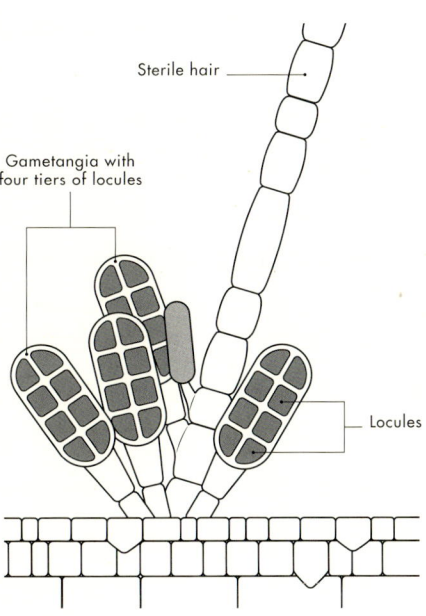

Plurilocular gametangia

Characteristic of the brown algae, microscopic plurilocular gametangia are divided into tiers of compartments called locules. Each locule contains one gamete.

Sterile hair

Gametangia with four tiers of locules

Locules

RED SEAWEED SEX

Sexual reproduction in the advanced red algae
(which make up 95 percent of all red algal species)
is remarkably uniform, but it differs markedly to the
process in all other algae. Male and female gametes in
the red algae lack flagella and are nonmotile, a notable
exception to the majority of algal gametes that have
flagella. The male gamete (the spermatium) of the red
algae is a microscopic spherical cell without flagella
formed in the male gametangia and released into the
sea when mature. It cannot swim and must be carried
by water movements to the female gamete. Also unique
to the advanced red algae, the female gamete (the
carpogonium) is the last cell in the special four-cell
carpogonial branch that forms on the gametophyte at
the onset of fertility. Clearly different to the vegetative
cells of the gametophyte, the carpogonium is a
flask-shaped cell with an inflated base and a long
tubular extension (the trichogyne) that extends beyond
the thallus. The carpogonium is not released from the
parental thallus as gametes typically are but is retained
on the gametophyte. The fertilized female gamete or
zygote is also retained on the gametophyte, where it
develops into another unique reproductive structure,
the carposporophyte. This structure forms spherical
spores (carpospores).

Red algal species also have sporophytes, which
again, remarkably, always produce four spherical spores
(tetraspores, "tetra" meaning four) in each sporangium.
The carpospores and tetraspores also lack flagella, in
marked contrast to the spores (zoospores or zoids)
of the green and brown seaweeds that typically have
flagella. The advanced red algae have three different
types of reproductive structures, which produce three
different types of reproductive cells: gametes,
carpospores, and tetraspores.

→ Microscopic whitish, urnlike
structures borne on the comblike
female thallus of the red seaweed
Herposiphonia pectinata are the
carposporophyte, the third phase
in the red algal life history.

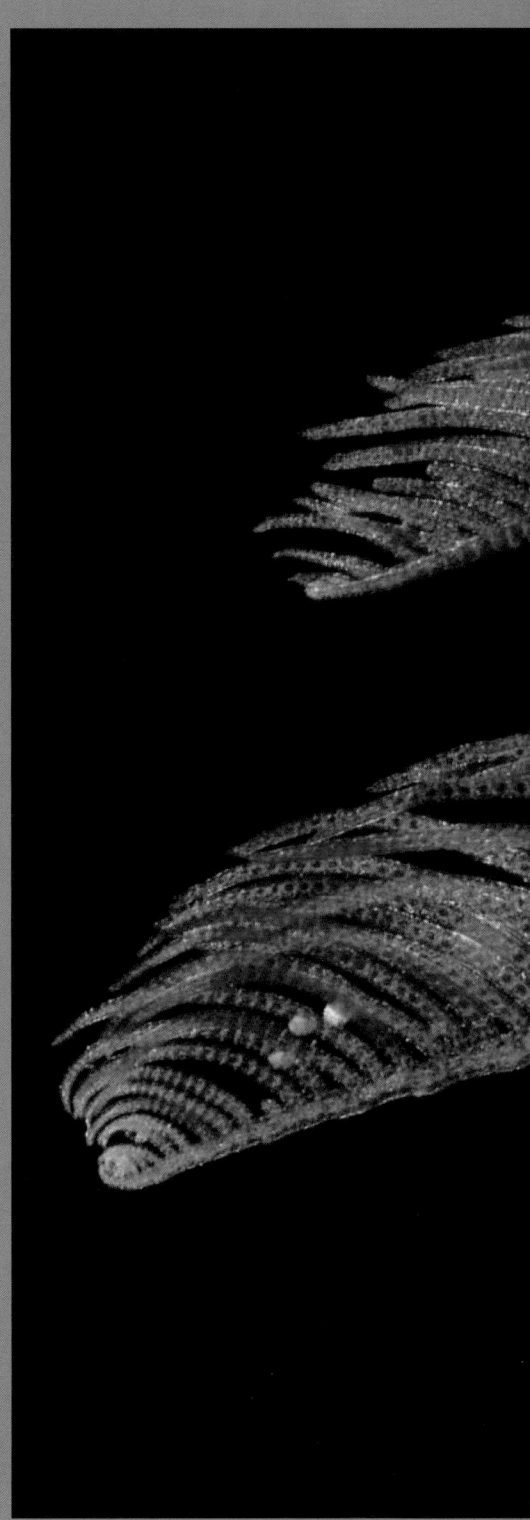

Three life history patterns

The great diversity in algal life histories is due to the variation in sexual reproduction in the algal lineages and the occurrence of meiosis relative to fertilization. Despite this diversity, three basic life history patterns are recognized. Two life history patterns—direct life histories—have thalli that produce only one type of reproductive cells (gametes) while the third pattern—the alternation of generations—has two different thallus types that produce either gametes or spores.

MEIOSIS AND FERTILIZATION

Meiosis and fertilization are the two key events in sexual reproduction. These events are responsible for the regular alternation of the two nuclear phases in the life history: the diploid phase (2n), with two sets of chromosomes, and the haploid phase (n), with one set of chromosomes. Meiosis is the nuclear division that reduces the chromosomes from the diploid to the haploid number, whereas fertilization is the process by which two haploid gametes fuse to form a diploid zygote. Fertilization results in the reestablishment of the diploid chromosome number.

Direct life histories have a single free-living phase, a thallus that can be either the cells of unicellular species or a multicellular macroalga. Gametes are the only reproductive cells formed by thalli with direct life histories. The two types of direct life histories are distinguished from one another depending on whether meiosis occurs either during the formation of gametes (a direct life history with gametic meiosis) or in the zygote (a direct life history with zygotic meiosis). The third life history pattern, "the alternation of generations," has two independent free-living generations: the gametophyte that produces gametes and the sporophyte that produces spores.

DIRECT LIFE HISTORY WITH ZYGOTIC MEIOSIS

In the direct life history with meiosis occurring in the zygote, cells of the unicellular alga or macroalga are haploid and form gametes by mitotic divisions. The gametes fuse to form diploid zygotes that subsequently divide by meiosis before germinating into a new haploid thallus. Meiosis is described as zygotic because it occurs in the zygote, the sole diploid stage in this life history pattern. Species of the unicellular green alga *Chlamydomonas* (phylum Chlorophyta), and the unicellular desmids and the multicellular *Chara* (see page 146) and *Spirogyra* in the green algal phylum Charophyta, have this life history pattern.

CHLAMYDOMONAS SEX

Life histories are known for only a few species of *Chlamydomonas*, including in great detail for the much studied *Chlamydomonas reinhardtii*. The opposite mating types (or opposite sexes) of the gametes of this species are the same size and morphology (isogamy). The onset of sex is marked by the flagellar tips of each gamete pair of the two mating types (mt+ and mt-) making the first contact. This is followed by adhesion along the length of the flagella, a process that entwines and eventually pairs the gametes. The pairing results in the apical region of the two gametes being brought into close contact, at which point a mating tube is produced by each cell. The protoplast of one or both cells traverse the mating tube and fuse (fertilization) to form a diploid zygote with four flagella, two from each gamete. Shortly after fertilization, the zygotes shed their flagella and develop a resistant, thick, spiny wall. After a period of dormancy, the zygote divides by meiosis into four haploid *Chlamydomonas* cells. In addition to the isogamous species, there are other species of *Chlamydomonas* that have either gametes

of unequal size (anisogamy) and eggs and spermatozoids (oogamy). The majority of cells in populations of *Chlamydomonas* divide asexually, with only relatively few cells acting as gametes during sexual episodes. The repeated asexual divisions enable these species to build up large populations in freshwater habitats.

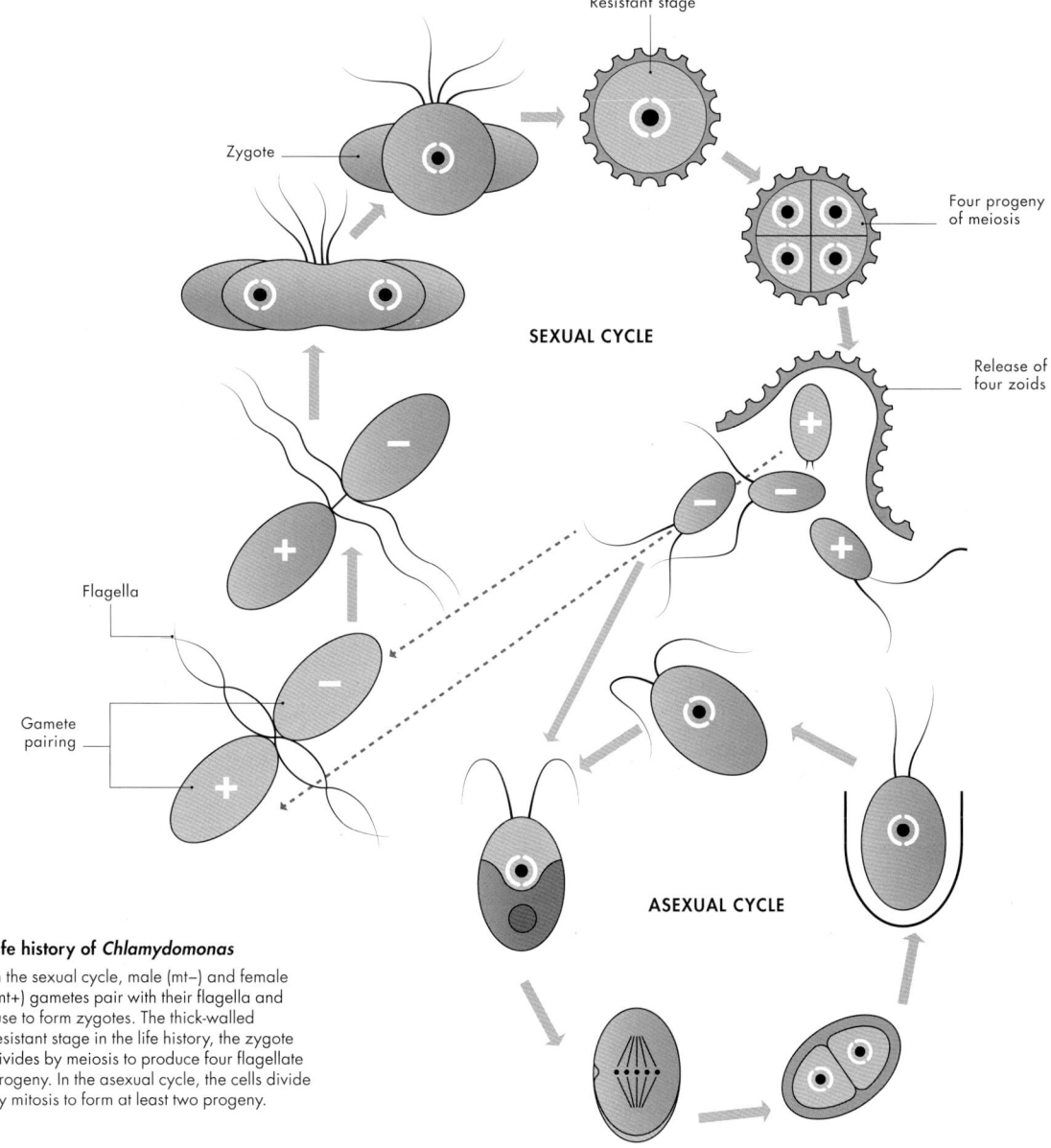

Life history of *Chlamydomonas*

In the sexual cycle, male (mt−) and female (mt+) gametes pair with their flagella and fuse to form zygotes. The thick-walled resistant stage in the life history, the zygote divides by meiosis to produce four flagellate progeny. In the asexual cycle, the cells divide by mitosis to form at least two progeny.

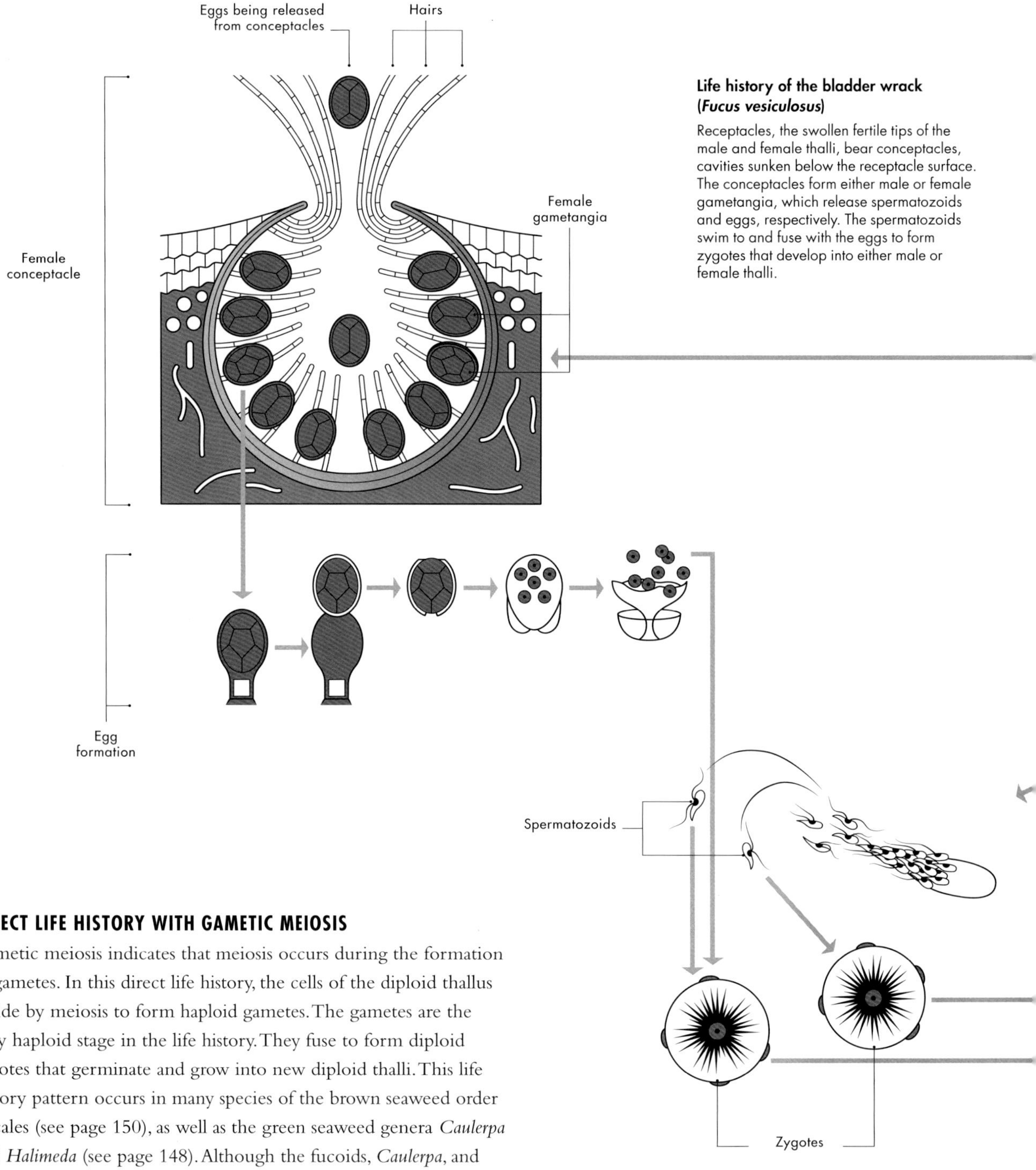

Eggs being released
from conceptacles

Hairs

Female
conceptacle

Female
gametangia

**Life history of the bladder wrack
(*Fucus vesiculosus*)**

Receptacles, the swollen fertile tips of the
male and female thalli, bear conceptacles,
cavities sunken below the receptacle surface.
The conceptacles form either male or female
gametangia, which release spermatozoids
and eggs, respectively. The spermatozoids
swim to and fuse with the eggs to form
zygotes that develop into either male or
female thalli.

Egg
formation

Spermatozoids

Zygotes

DIRECT LIFE HISTORY WITH GAMETIC MEIOSIS

Gametic meiosis indicates that meiosis occurs during the formation
of gametes. In this direct life history, the cells of the diploid thallus
divide by meiosis to form haploid gametes. The gametes are the
only haploid stage in the life history. They fuse to form diploid
zygotes that germinate and grow into new diploid thalli. This life
history pattern occurs in many species of the brown seaweed order
Fucales (see page 150), as well as the green seaweed genera *Caulerpa*
and *Halimeda* (see page 148). Although the fucoids, *Caulerpa*, and
Halimeda have the same direct life history pattern, the structure
of their reproductive organs and gametes are vastly different. For
example, the gametes of the fucoids are eggs and spermatozoids,
while gametes released by the thalli of *Caulerpa* and *Halimeda* are
isogametes and have the same morphology and size.

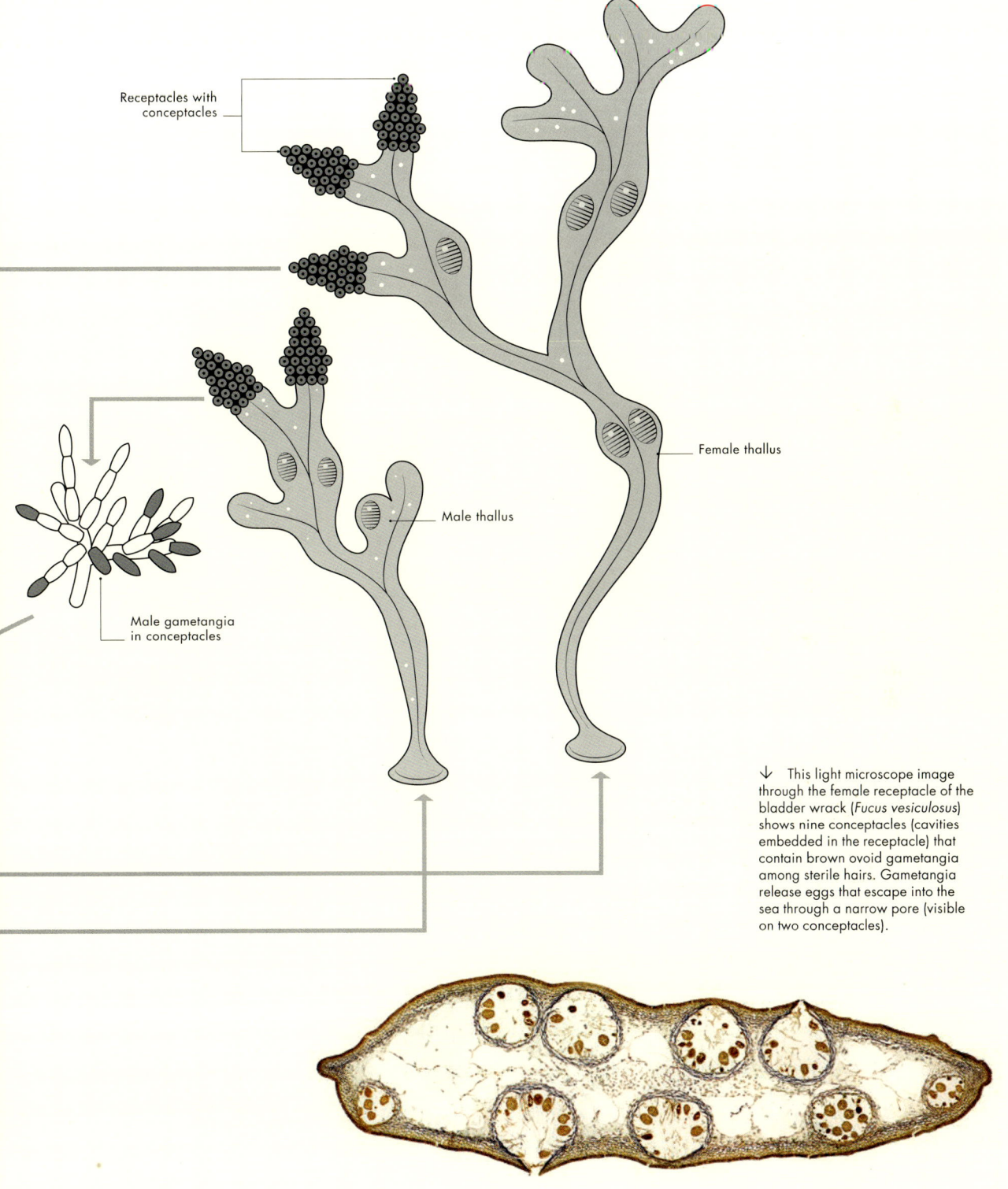

Receptacles with
conceptacles

Male gametangia
in conceptacles

Male thallus

Female thallus

↓   This light microscope image
through the female receptacle of the
bladder wrack (*Fucus vesiculosus*)
shows nine conceptacles (cavities
embedded in the receptacle) that
contain brown ovoid gametangia
among sterile hairs. Gametangia
release eggs that escape into the
sea through a narrow pore (visible
on two conceptacles).

ALTERNATION OF GENERATIONS

In the third life history pattern—the alternation of generations—the haploid gametophyte generation alternates with the diploid sporophyte generation. The two major events in sexual reproduction—fertilization and meiosis—occur in the different generations. Fertilization results from the fusion of the gametes that are produced by the gametophytes and meiosis occurs in the sporophytes. This life history pattern is subdivided into two types based on whether the gametophytes and sporophytes look alike (isomorphic generations) or very different (heteromorphic generations).

ISOMORPHIC GENERATIONS

Species of the green seaweeds *Ulva* (see page 150) and *Cladophora*, and the brown seaweeds *Ectocarpus* and *Dictyota* (see page 152) have life histories that involve an alternation of isomorphic generations. The gametophytes and sporophytes of these species look alike but can be distinguished by counting their chromosomes. Sporophytes are diploid and have double the number of chromosomes found in the gametophytes of the same species. They may also be distinguished by the reproductive cells they produce or their reproductive organs. Gametophytes of *Ulva* and *Cladophora* produce gametes with two flagella

Life history of the sea lettuce (*Ulva*)

Biflagellate male and female gametes produced by male (−) and female (+) gametophytes (left) fuse to form zygotes which, in turn, grow into diploid sporophytes. Meiosis in the sporophyte produces quadriflagellate haploid zoospores, which re-establish the gametophyte generation.

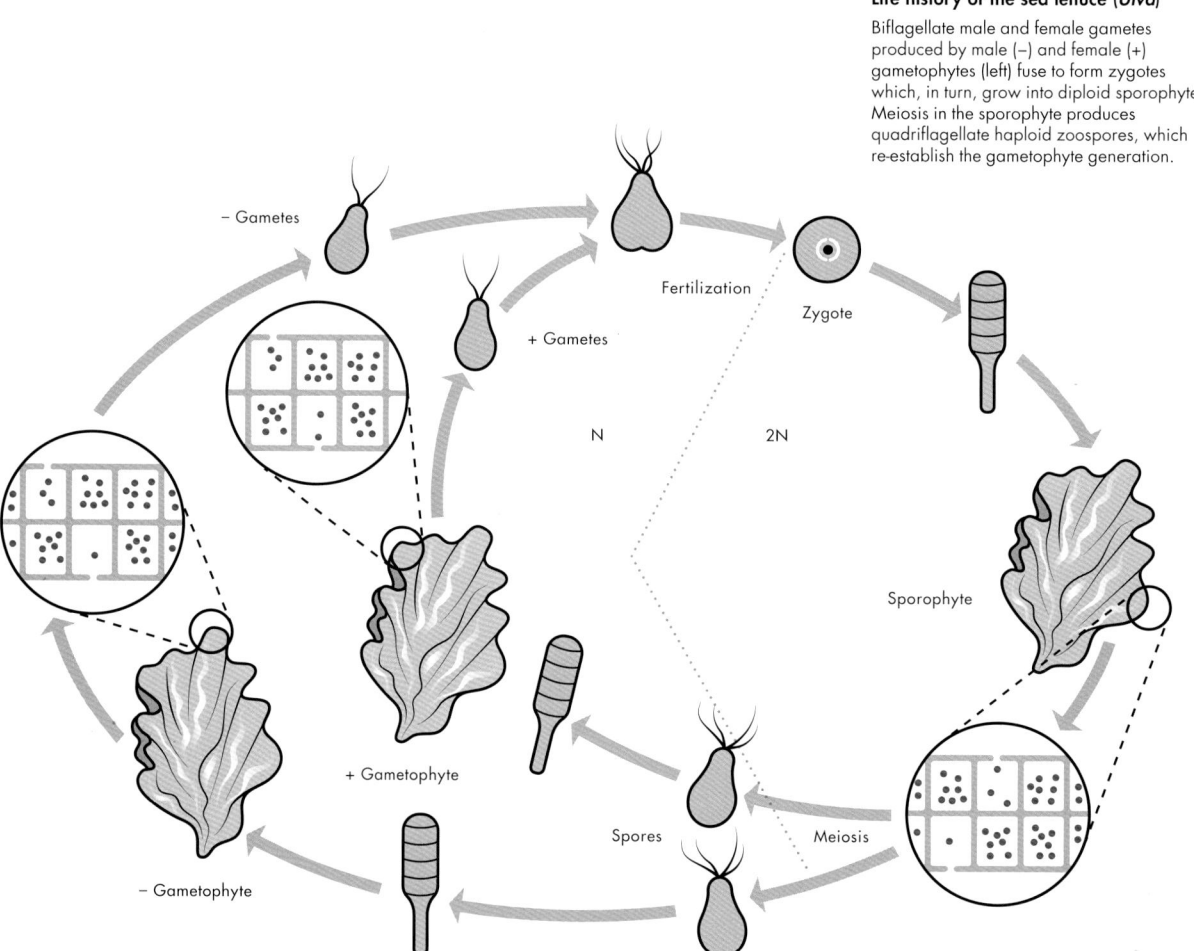

− Gametes

+ Gametes

Fertilization

Zygote

N

2N

Sporophyte

+ Gametophyte

− Gametophyte

Spores

Meiosis

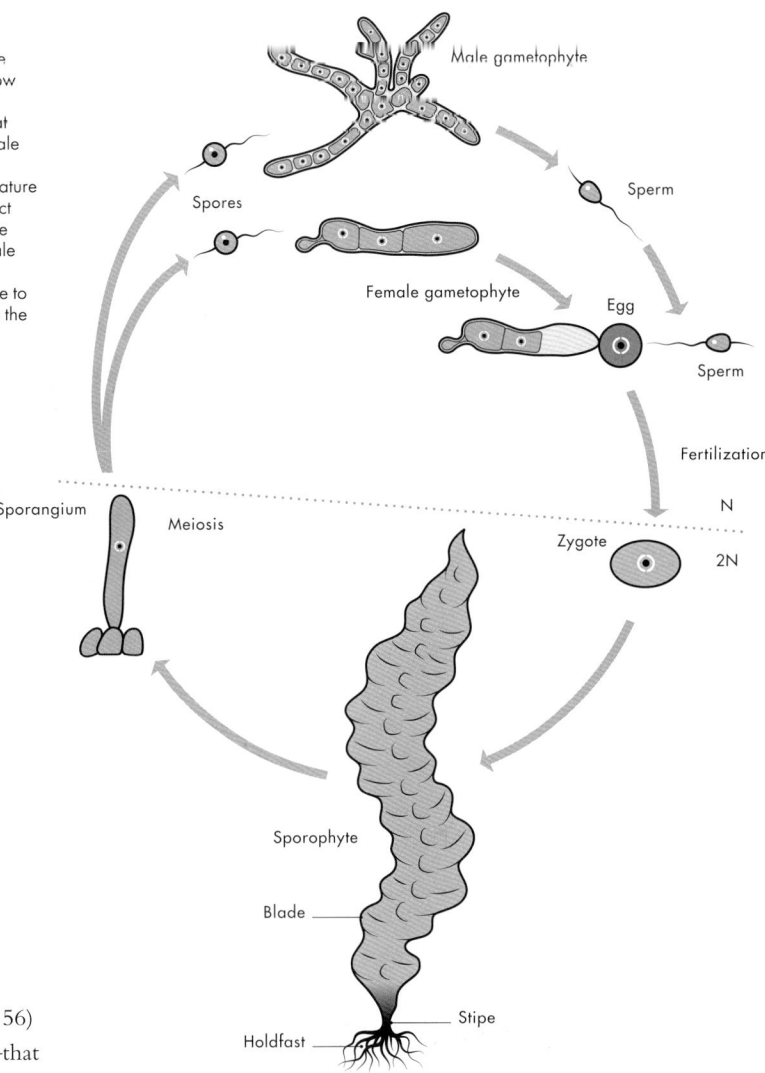

Life history of a kelp

Meiosis in the sporangia of the large diploid kelp thallus (below right) results in the release of haploid biflagellate spores that germinate into microscopic male and female gametophytes. Spermatozoids released by mature male gametophytes first contact with their anterior flagellum the eggs released by mature female gametophytes. Eggs and spermatozoids fuse, giving rise to diploid zygotes that grow into the diploid kelp thallus.

and the sporophytes produce zoospores (named for their ability to swim like unicellular animals) with four flagella. In *Ectocarpus*, the gametangia on the gametophytes are divided into many tiers of locules, whereas the sporangia on the sporophytes are composed of one large locule.

HETEROMORPHIC GENERATIONS

In an alternation of heteromorphic generations, the gametophytes and sporophytes have a different appearance. One generation is a large thallus—the macrothallus—while the other generation has a microscopic thallus. The familiar kelps (see page 156) are the sporophyte generation—the macrothallus—that alternates with microscopic gametophytes that often are composed of several cells. In the green seaweed *Bryopsis plumosa* the erect, featherlike gametophyte—the macrothallus—grows to 4 in (10 cm) in height and alternates with a small, creeping sporophyte. This life history type is an adaptation to an environment that is seasonality variable, or an avoidance to herbivory. In species that grow on rocky shores above low-tide mark, the macrothallus is often destroyed during hot summers, whereas the microthallus inhabits a more favorable habitat in crevices and small pools of seawater. The macrothallus is more easily grazed by herbivores than the microthallus.

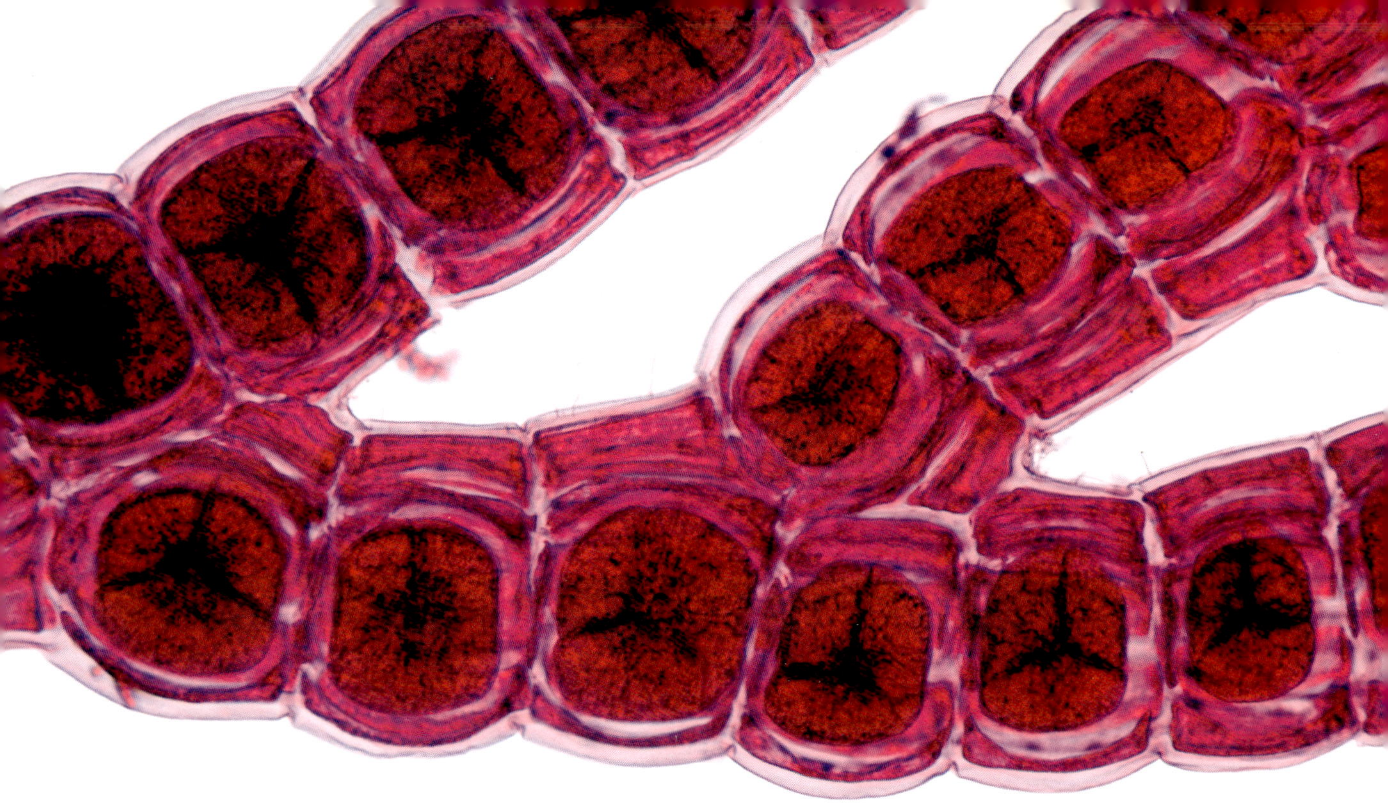

THE RED ALGAL LIFE HISTORY

A life history with an alternation of generations has been modified into a triphasic life history that is unique to the advanced red algae. The three phases in the life history are the free-living sporophytes and gametophytes that are found in algal species with an alternation of generations, plus an additional sporophyte phase (the carposporophyte) that grows on the female gametophyte.

In this red algal life history, the haploid gametes are generated by mitosis in the haploid cells of the gametophytes. The female gamete (the carpogonium) is not released into the sea, which typically happens in the algae, but is retained on the gametophyte. The male gamete (the spermatium) that lacks flagella is released into the sea where it passively drifts until it is "caught" by the long tubular extension (the trichogyne) of the female gamete. The male gamete injects its nucleus into the tube. The long tube is thought to be an adaptation to increase the chances of fertilization by male gametes that cannot swim to the female gamete. The nucleus travels down the tube and fertilizes the female gamete. What ensues next is a series of complex postfertilization events that culminate in the formation of the diploid carposporophyte. These events provide the characters for classifying species in the various orders of the red algae.

Mitotic divisions in the carposporophyte generate many diploid spherical carpospores that are released into the sea. The carpospores germinate into diploid sporophytes which, when mature, form sporangia. These sporangia divide by meiosis to form four haploid

spores that grow into the gametophyte generation.
Depending on the species, the carpospores range in
size from 17 to 121.9 microns in diameter, while the
tetraspores range from 3 to 182.5 microns.

DINOFLAGELLATE LIFE HISTORY

Recent studies have revealed that the dinoflagellate
life history is more varied and flexible than previously
believed. In favorable environments dinoflagellates
reproduce asexually by mitosis, which maintains the
population size. Asexual reproduction was thought to
be the dominant mode of reproduction.

However, sexual reproduction is now known to be
a cryptic process. It is difficult to observe because the
gametes resemble the vegetative cells; gamete fusion is
a very slow process that is difficult to distinguish from
cell division and some species reproduce sexually at night.
Sexual reproduction is now thought to be a more regular
event, with many species considered to have a direct life
history with meiosis occurring in the zygote.

The dinoflagellate vegetative cell is haploid. The male and
female gametes have the same morphology but, depending
on the species, may be the same size (isogamy), or the
female is larger than the male gamete (anisogamy). There
is often a mating reaction with pairs of gametes dancing
around each other and/or coupling through their flagella.
The gametes fuse and the protoplast of one is transferred
into the protoplast of the other, forming a diploid zygote.
The zygote is assumed to divide by meiosis, becoming
either a vegetative cell or a resistant stage in the life history.

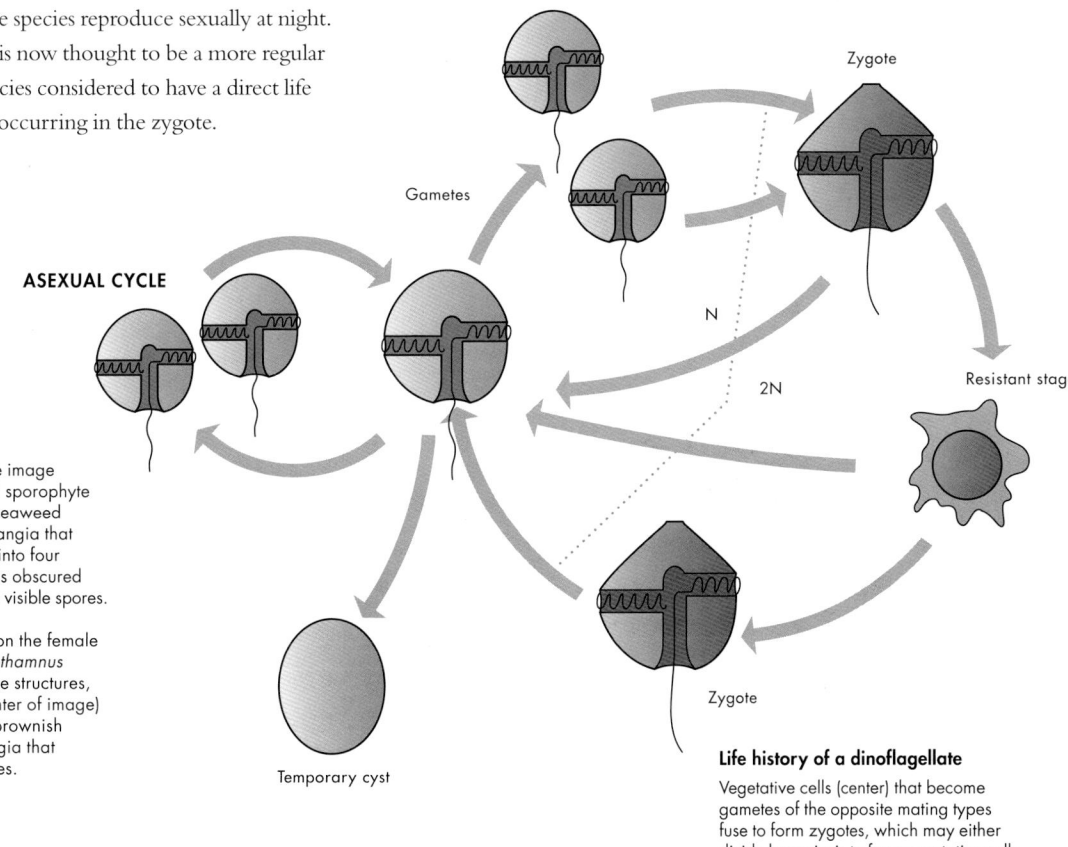

↖ This light microscope image
of the filamentous diploid sporophyte
(stained pink) of the red seaweed
Polysiphonia shows sporangia that
have divided by meiosis into four
spores. The fourth spore is obscured
from view behind the three visible spores.

← Carposporophytes on the female
gametophytes of *Melanothamnus
sphaerocarpus* are urnlike structures,
three of which (lower center of image)
have formed elongated brownish
(unstained) carposporangia that
contain many carpospores.

Life history of a dinoflagellate
Vegetative cells (center) that become
gametes of the opposite mating types
fuse to form zygotes, which may either
divide by meiosis to form vegetative cells
or develop into a dormant resistant
stage. Vegetative cells may also form
temporary cysts or divide asexually
by mitosis.

DIATOM LIFE HISTORY

The diatom life history also involves both asexual and sexual reproduction. Diatom cells reproduce asexually by mitosis and, without the intervention of sex, repeated mitotic divisions result in a progressive reduction of cell size. Sexual reproduction in the diatoms has the additional function of restoring cell size. The diatoms have a direct life history with meiosis occurring during gamete formation. The diploid vegetative cells of the diatoms form the gametes. The two main groups of diatoms—the centric and pennate diatoms—have different aspects to their life histories.

Life history of a centric diatom
A uniflagellate spermatozoid released by the male diatom cell swims to and fertilizes the female diatom cell (center left) to form a zygote, known as an auxospore. It functions to restore the size of the vegetative cells following their diminution by repeated mitotic divisions, evident in the four cells in lower center and the two cells at lower left.

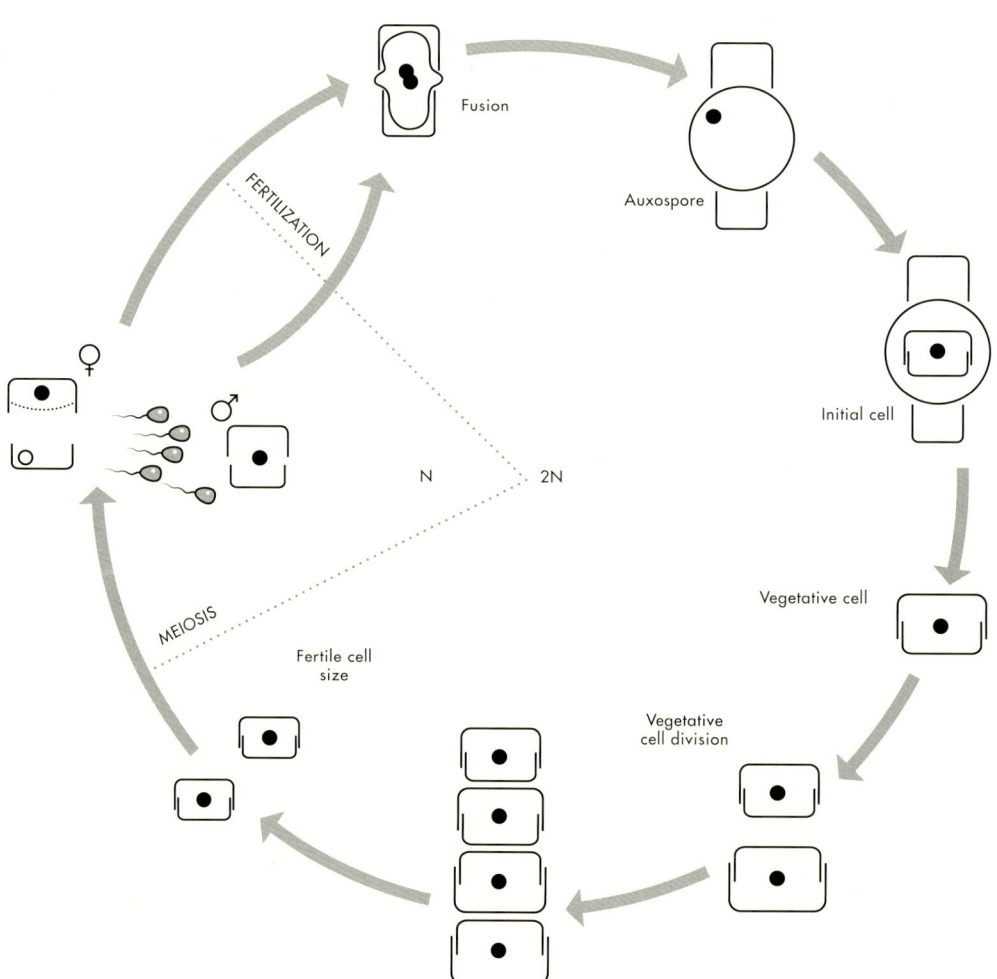

← A series of light microscope images shows stages in sexual reproduction of the pennate diatom *Sellaphora auldreekie*, whose vegetative cell (1) has a single brown H-shaped plastid. Mating begins with the side-by-side pairing of cells (2). During early meiosis, the protoplast reorganizes through movements of the plastid and nucleus (the large transparent sphere in the cell center) (3). One large gamete forms in each cell; the cell walls break down at the point of contact to form an aperture (4) through which the male gamete moves to fertilize the female gamete (5). The zygote enlarges to form a cigar-shaped auxospore (6).

SEX IN CENTRIC DIATOMS

The female parent cell of a centric diatom produces one or two egg cells, the other two or three progeny of the meiotic division dying. The male parent cell produces from four to 128 uniflagellate spermatozoids initially by meiotic division, followed by a variable number of mitotic divisions (in cells with more than four spermatozoids). When the gametes are mature, the frustules open between the valves, allowing the spermatozoids to escape. The spermatozoids swim to fuse with the egg, forming the zygote (known as the auxospore). The auxospore has the special function of reinstituting cell size by shaping new, much larger frustules—the rigid cell wall that encloses the diatom cell. It achieves this by increasing in size and forming a much larger organic cell wall covered with silica scales. The new frustule develops inside the organic cell wall.

SEX IN PENNATE DIATOMS

Sexual reproduction in pennate diatoms begins with the pairing of diploid parent cells enclosed in mucilage. Meiosis in the parent cell results in only one haploid gamete, the other three nuclei die during the division. The gametes which emerge when the frustules gape open are similar (isogamy) and lack flagella. One active gamete moves to fuse with the passive gamete to form the zygote (auxospore), which greatly increases in size to restore the cell size lost during mitotic divisions.

↑ The microscopic carposporophyte of the red seaweed *Asparagopsis armata* releases a spherical mass comprising hundreds of comparatively small carposporangia.

↗ A series of tetrasporangia of the filamentous red seaweed *Platysiphonia delicata* form as two longitudinal rows on the short lateral branches of the plant.

Resistant stages

The ability to form resistant, resting, or dormant stages is a common trait in the life history of species from many algal phyla. Dinoflagellates, diatoms, and freshwater green algae form resistant stages but the red and brown seaweeds do not. Resistant stages may provide both the means for survival during adverse environmental conditions and the capacity for a rapid increase in population size when the environment becomes favorable.

DINOFLAGELLATE RESISTANT STAGES

Many dinoflagellate species form resistant cysts with an impervious organic cell wall. These cysts are resistant to extreme environmental conditions, evident from the fossil cysts preserved in sedimentary rocks for more than 100 million years. Fossil dinoflagellate cysts are used in petroleum exploration to identify oil-bearing rock layers. Approximately 150 species of living dinoflagellates are known to form cysts, which may be spherical, or resemble the motile cells of the same species, or have elaborate extensions on the cyst wall. When the environment becomes favorable, dormant cysts in the sediments germinate and one dinoflagellate cell exits from each cyst to resume its pelagic life. The mass germination of dinoflagellate cysts may lead to the rapid increase in population size and the formation of dinoflagellate blooms.

Fossil dinoflagellate cysts

A moderate number of long processes with branched tips arise from the thecal plates that cover the cell of this microscopic fossil cyst.

DIATOM RESTING STAGES

Resting cells or spores are produced by certain
species in many diatom genera, including *Chaetoceros*,
Asterionella, *Tabellaria*, and *Fragilaria*, but they are not
as long-lived as the dinoflagellate cysts. The resting
stages may be morphologically similar to the diatom
vegetative cell, or they may have a very thick, rounded,
and less ornamented frustule. These structures are
loaded with nutrients and may lay dormant in the
sediments for decades.

RESISTANT STAGE OF THE GREEN ALGAE

Species of the green alga *Chara* and its close relatives
form thick-walled zygotes with an impervious layer
in their cell wall, while many of the unicellular
chlorophytan green algae found living in unpredictable
aquatic habitats, or harsh terrestrial environments, form
resistant cysts during periods of adverse environmental
conditions. Under extreme environmental conditions,
Haematococcus lacustris—a unicellular green alga with
the intriguing common name of the "blood-red
alga"—forms cysts with a wall that is highly resistant
to chemical, enzymatic, and mechanical attack.

Dinoflagellate motile cells and cysts
Microscopic motile cells of
Protoperidinium subinerme (1) and
Protoperidinium pentagonium (4),
with a cell covering of thecal plates,
are somewhat similar in shape to their
cysts: *Protoperidinium subinerme* (2);
Protoperidinium pentagonium (3).

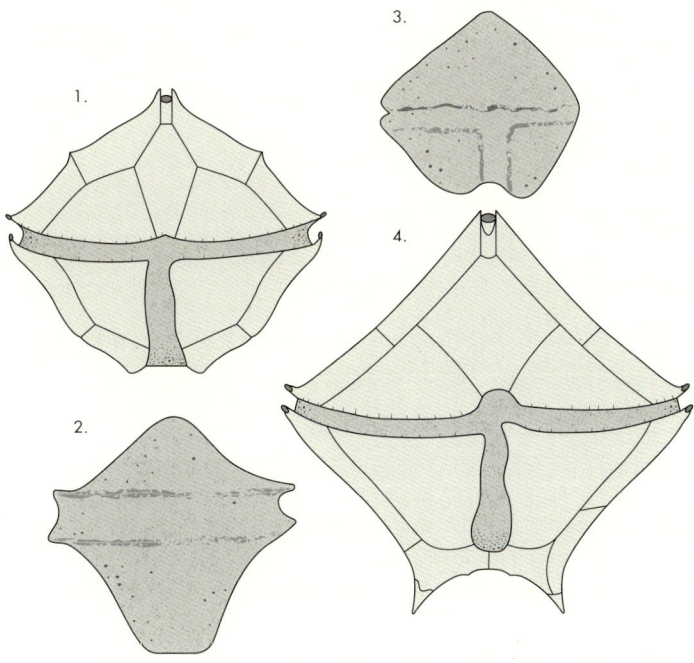

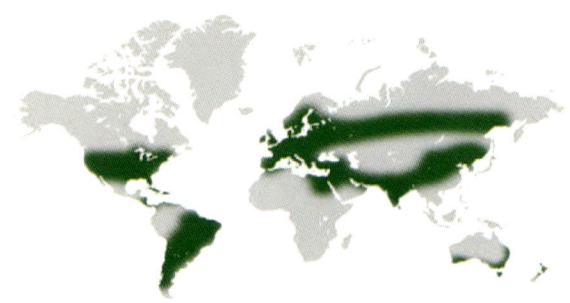

KINGDOM	Plantae
PHYLUM	Charophyta
CLASS	Charophyceae
ORDER	Charales
GENUS	*Chara*
SIZE	Oogonium to 900 microns long; antheridia to 500 microns in diameter
HABITAT	Freshwater and brackish ponds, and rivers

PHYLUM CHAROPHYTA

Chara vulgaris

Common stonewort

Chara vulgaris has only one phase in the life history, the gametophyte, that may grow to 20 in (around 50 cm) tall. Unique among the green algae, species of Chara and their close relatives have multicellular gametangia which, when mature, are visible to the unaided eye.

DIRECT LIFE HISTORY

Gametes form in gametangia, which are the only reproductive structures formed on plants of *Chara vulgaris*. In this species, the male and female gametangia are borne of the same thallus among the whorls of lateral branches. The gametangia are composed of different cell types, a feature that is not only unique in the algae to *Chara* and its very close relatives, but also aligns them with the primitive land plants.

The spherical male gametangium is composed of three different cell types. The outermost shield cells contain orange carotenoid droplets that color the gametangium bright orange and are arranged in an intriguing flowerlike array of eight cells. Cells derived from another cell type (the capitulum) form long unbranched filaments that produce a single spermatozoid. Each male gametangium releases thousands of spermatozoids, which resemble those of the primitive land plants rather than other green algal motile cells.

The brown cylindrical female gametangium is protected by five sheath cells that spiral clockwise around the single egg. Each sheath cell divides to form a crown cell. The spermatozoid enters the female gametangium through the fissures between the crown and sheath cells at the top of the gametangium and fertilizes the egg.

After fertilization, rapid changes in the sheath cells result in the formation of an eight-layered darkly pigmented wall around the zygote. The outermost ornamentation layer contains highly resistant organic material thought to protect the zygote from desiccation and grazing. The zygote falls into the sediments of freshwater habitats and remains dormant there until it divides by meiosis, germinates and grows into a new *Chara* thallus.

The spermatozoids and the vegetative cells of *Chara* have the same quantity of DNA, which provides evidence that meiosis occurs inside the thick walled zygote. If meiosis occurred during the formation of gametes, the spermatozoids would have half the quantity of DNA found in the vegetative cells of *Chara*.

→   Light microscope image of the common stonewort (*Chara vulgaris*) shows the brown female gametangium (above), with spiralled sheath cells capped by crown cells, and the bright orange male gametangium (below), with its petallike array of shield cells, among the very short lateral branches on the nodes of the main axis.

PHYLUM CHLOROPHYTA

Halimeda incrassata

Siphonous green alga

KINGDOM	Plantae
PHYLUM	Chlorophyta
CLASS	Ulvophyceae
ORDER	Bryopsidales
GENUS	*Halimeda*
SIZE	Female gamete 13.5–19.5 microns long
HABITAT	Grows on sand or rock on coasts and coralgal reefs to depths of 215 ft (65 m)

During sexual episodes, species of *Halimeda* have adapted to their thallus structure—an aggregation of long tubes (siphons)—by converting the entire protoplast in the thallus into gametes. There is no structure in the *Halimeda* thallus in which the gametes could be segregated from the rest of the thallus.

The *Halimeda* plant dies after the gametes are released into the sea at its first and only reproductive episode. Therefore, unsurprisingly, sexual reproduction in this genus is comparatively rare. Only around 5 percent of individuals in a population reproduce during each sexual episode. Nevertheless, these seasonal, synchronous, short-lived bouts of sexual reproduction are highly organized and a remarkable feature of coralgal reefs. In these habitats, the large populations of *Halimeda* are maintained by asexual reproduction that generates new plants through elongating rhizoids and thallus fragments.

DIRECT LIFE HISTORY

Halimeda incrassata has a direct life history. The onset of sexual reproduction is marked by the transfer of the entire protoplast of the thallus into simple gametangia that are merely outgrowths of the siphons on the margins of each thallus segment. This process turns the thallus a limey white and forms (just visible to the unaided eye) gametangia that resemble bundles of dark-green grapes fringing the thallus margins (see page 125). This occurs 23–24 hours before gamete release,

giving the human observer one day's advance notice of impending and potentially easily missed sexual events. The protoplast divides by meiosis to produce biflagellate gametes that are released around dawn in a concert of brief, massive pulses of reproductive activity that lasts from 5 to 20 minutes. The released gametes form copious green clouds that stream across the reef for 15 minutes before dissipating.

The gametes fuse to form zygotes, some of which are dispersed across the reef and grow into new diploid *Halimeda* thalli. Constrained by a siphonous thallus that lacks any partitioned compartment in which gametes could form, *Halimeda* and its entirely siphonous relatives have holocarpic reproduction, a reproductive pattern unique to siphonous green algae and defined as converting the entire thallus protoplast into gametes. Their release results in the death of the plants, which leaves empty, white "ghost" thalli in its wake. The dead *Halimeda* thalli fall apart and their loose calcareous segments make a significant contribution to the sediments and the formation of limestone (carbonate rock) on coralgal reefs.

→ At its first and only sexual episode, the entire protoplast of the siphonous segmented thallus of *Halimeda incrassata* is converted into gametes. Their release into the sea results in the death of the parent plant.

KINGDOM : Chromista
PHYLUM : Ochrophyta
CLASS : Phaeophyceae
ORDER : Fucales
GENUS : *Ascophyllum*
SIZE : Fucoid eggs 80 microns in diameter
HABITAT : Intertidal zone, sheltered rocky seashores, and estuaries

CLASS PHAEOPHYCEAE

Ascophyllum nodosum

Knotted wrack

The knotted wrack (*Ascophyllum nodosum*) is a long-lived perennial that sometimes survives for 30–40 years. This species has a seasonal pattern of growth and reproduction. In early spring, the onset of vegetative growth is marked by the development on each branch, behind the branch apex, of a pneumatocyst measuring 0.4–0.6 in (1–1.5 cm) in diameter.

The reproductive structures of the knotted wrack form in receptacles. These are special fertile branches that are initiated each year before the fertile period. In summer, the portions of the thallus from the previous year's growth, which are immediately behind the newly formed pneumatocysts as well as the older thallus portions, initiate the development of many receptacles. These continue to develop for almost a year, culminating in the release of gametes the following April and May.

DIRECT LIFE HISTORY

The knotted wrack has a direct life history. The only free living phase in the life history, the large 3.3–9.9 ft (1–3 m) thallus is diploid and produces gametes by meiotic divisions. Each year, in February and March, the short days (less than 10 hours of light) initiate the development of the reproductive structures, the conceptacles, in the receptacles that formed during the previous summer. The conceptacles form from

an infolding of the outermost cell layer of the thallus (the meristoderm), which creates a spherical cavity buried in the thallus. Each conceptacle opens onto the thallus surface via a short narrow passage and a pore. Many conceptacles form in the receptacles.

Numerous female gametangia that form in the conceptacles on the female thallus divide by meiosis to produce four eggs in each gametangium. Meiotic divisions that also occur in the numerous male gametangia in the conceptacles on the male thallus generate large numbers of spermatozoids. The eggs and spermatozoids are released into the sea through the pore of their conceptacles. Once in the sea, the eggs secrete the chemical sex attractant finavarrene, which attracts the spermatozoids to swim to the eggs and fertilize them. After fertilization the zygotes sink, settle on the substratum, and germinate into new thalli.

→ Fertile clublike receptacles of the knotted wrack (*Ascophyllum nodosum*), which form behind the pneumatocyst (top center) on either side of the main thallus branches, have a mottled appearance from the many conceptacles—flask-shaped cavities containing gametangia within the receptacle.

KINGDOM	:	Plantae
PHYLUM	:	Chlorophyta
CLASS	:	Ulvophyceae
ORDER	:	Ulvales
GENUS	:	*Ulva*
SIZE	:	Female gamete 7–11 microns long, 4–6 microns across
HABITAT	:	Intertidal zone on rocky coasts

PHYLUM CHLOROPHYTA

Ulva rigida

Sea lettuce

Commonly known as the sea lettuces, *Ulva* plants grow as small tufts to larger blades 12 in (30 cm) or more tall. The life history of the sea lettuce is an alternation of isomorphic generations, defined as having sporophytes and gametophytes that look alike.

ISOMORPHIC GENERATIONS

Sea lettuces lack specialized reproductive structures (see page 134). Instead, the vegetative cells on the upper thallus margins simply divide by mitosis or meiosis to form the reproductive cells. These cells on the haploid male and female gametophytes divide by mitosis to form 16 or more biflagellate gametes in each cell. The gametes that are released into the sea at dawn swim, directed by their light detecting eyespot, toward the light. This maximizes the chance of fertilization by forming swarms at the sea surface in which the male and female gametes fuse to form diploid zygotes. The zygotes swim to and settle on a rock and grow into diploid sporophytes.

Cells of the sporophyte divide by meiosis to form 8 or 16 haploid zoospores, which have four flagella. The zoospores swim to and settle on the rocky sea bottom where they grow into gametophytes. Any gametes of the sea lettuces that do not fuse with another gamete develop into new thalli, a testament to the weediness of the sea lettuces when every gamete, fertilized or not, can grow into a new plant.

ECOLOGICAL SUCCESS

Mature thalli of sea lettuces release gametes and zoospores at least at two-weekly intervals on the spring tides, the highest high and lowest low tides of the month. These frequent reproductive episodes, coupled with the capacity for all zoospores, unfused gametes, and zygotes to germinate into new plants, provide a constant supply of reproductive cells to rapidly colonize any available space created by ecological disturbance on the seashore.

The life history traits of the sea lettuces coupled with the tolerance of a wide range of environmental conditions, and the ability to grow rapidly in high-nutrient environments, contribute to the ability of these weedy opportunistic species to form massive algal blooms (see page 207).

→ Sexual reproduction in the sea lettuce *Ulva rigida* is a simple process: cells on the thallus margin of the sporophytes divide by meiosis to form zoospores, whereas those cells on the gametophytes form gametes by mitosis.

CLASS PHAEOPHYCEAE

Dictyota diemensis

Brown alga

KINGDOM	Chromista
PHYLUM	Ochrophyta
CLASS	Phaeophyceae
ORDER	Dictyotales
GENUS	*Dictyota*
SIZE	Ova 58–70 microns in diameter; spermatozoid cell 3–4 microns in diameter
HABITAT	Upper sublittoral zone on wave-exposed coasts

The life history of the brown seaweed *Dictyota diemensis* is an alternation of isomorphic generations, which is defined as having gametophytes and sporophytes that are similar in appearance.

The branches of the much branched thalli of this species repeatedly divide into two equal branches. The thalli grow to 12 in (30 cm) tall with each branch measuring 0.2–0.4 in (5–10 mm) wide. *Dictyota diemensis* is fertile from November to February, during late spring and summer in the southern hemisphere. For those four months, the gametophytes release gametes on two consecutive days once a month, a total of only eight days a year.

ISOMORPHIC GENERATIONS

The gametangia on the gametophytes of *Dictyota diemensis* are tightly packed into patchlike sori that are scattered on both the upper and lower thallus surfaces. The sori are visible to the unaided eye. The many white sori on the male gametophytes are composed of many white gametangia that are markedly different to the typically brown male gametangia of the brown seaweeds. Each male gametangium is divided into 16–26 tiers of compartments with 16 compartments in each tier. Each compartment contains one spermatozoid. The massive release of thousands of spermatozoids into the sea around dawn from the male gametophytes momentarily turns the seawater milky.

Unlike the white male sori, the sori on the female gametophytes are brown and have many female gametangia tightly packed in rows. Each female gametangium releases only one of the hundreds of eggs released by each female thallus. The eggs secrete the chemical sex attractant dictyotene into the seawater, which attracts the spermatozoids causing them to swim to the eggs. The spermatozoids cluster around an egg forming a halo, until one succeeds in fertilizing it.

The zygote germinates into a diploid sporophyte that, when mature, produces sporangial sori that are also scattered across both thalli surfaces. Many brown sporangia are loosely clustered into the sorus and are at different stages of development, two characters that distinguish a sporangial sorus from a female sorus. The sporangium divides by meiosis to form four spherical nonflagellate tetraspores, which range in diameter from 60 to 77 microns. *Dictyota* and its close relatives (in the order Dictyotales) are the only brown seaweeds to reproduce by tetraspores. The haploid tetraspores germinate and grow into gametophytes.

→ Dried herbarium specimen of *Dictyota diemensis*, a species whose gametophytes and sporophytes are morphologically indistinguishable until fertile, when the different specialized reproductive structures permit the identification of the three thallus types.

CLASS PHAEOPHYCEAE

Scytosiphon lomentaria

Brown alga

KINGDOM	Chromista
PHYLUM	Ochrophyta
CLASS	Phaeophyceae
ORDER	Ectocarpales
GENUS	*Scytosiphon*
SIZE	Female gamete 3.7–5.7 microns long; male gamete 3.7–4.4 microns long
HABITAT	Intertidal zone on rocky coasts

Scytosiphon lomentaria (a brown alga) has an alternation of heteromorphic generations. The macroscopic phase in the life history is a narrow, hollow, tubular brown alga standing 20–27.5 in (50–70 cm) tall, which alternates with a disklike, flat crust that may reach a maximum size of 2 in (50 mm) across and, when sterile, 60–100 microns thick.

HETEROMORPHIC GENERATIONS

As inhabitants of the harsh intertidal zone, both phases are subjected to a variety of stressors including desiccation, ultraviolet radiation, and intense heat or cold. Of the two phases, the crust is the resistant stage in the life history, being less susceptible to the harsh conditions of winter at high latitudes and summer at lower latitudes, as well as sand scour, burial, wave shear, and grazing by herbivores.

→   The macroscopic tubular thallus of *Scytosiphon lomentaria* is the gametophyte generation that alternates with the sporophyte generation, an environmentally and biologically resistant small filamentous crust.

In the sexual cycle, the tubular thalli of *Scytosiphon lomentaria* are the haploid male and female gametophytes that bear plurilocular gametangia. These are clustered into sori that form brown patches on the thallus surface. Female gametes liberated by the gametangia swim to and settle on a rock. Once settled, they secrete the chemical sex attractant hormosirene. A mating reaction ensues, with a swarm of male gametes clustering around the female gamete until fertilization occurs. The zygotes germinate into the diploid sporophyte, which, in this species, is a crust, a flat disk made of filaments that closely adhere to the rock. The crust develops unilocular sporangia that divide by meiosis to produce flagellate haploid zoids that germinate and grow into gametophyte generation.

"DEVIANT" LIFE HISTORY

The sexual cycle in *Scytosiphon lomentaria* has rarely been observed. More common is a simplified, entirely haploid "deviant" cycle based on the germination of unfused female and male gametes. The unfused gametes develop into haploid crusts that are different to the diploid crusts of the sexual cycle. New erect tubular thalli then develop on the haploid crust. Many populations of *Scytosiphon lomentaria* in temperate regions worldwide have been reported to reproduce solely by the haploid cycle and thus are asexual populations, thought to have lost the ability to reproduce sexually.

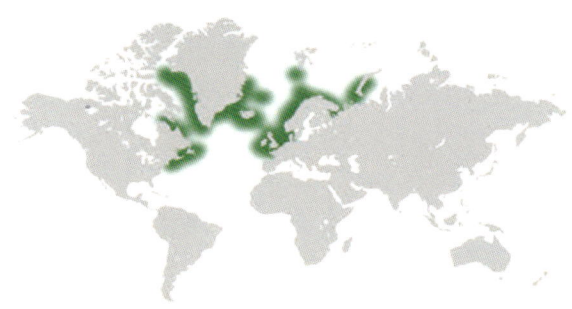

KINGDOM	Chromista
PHYLUM	Ochrophyta
CLASS	Phaeophyceae
ORDER	Laminariales
GENUS	*Laminaria*
SIZE	Ovum 15-30 microns in diameter; spermatozoid cell 6–7 microns long
HABITAT	Lower intertidal to upper subtidal zone

CLASS PHAEOPHYCEAE

Laminaria digitata

Oarweed

The familiar 6.5 ft (2 m) leathery blades of the kelp, commonly called oarweed (*Laminaria digitata*), are the macrothallus (diploid sporophyte), which alternates with a microscopic filamentous microthallus (haploid gametophyte).

Sporangial sori form on the kelp blades from April to December, and zoids are released from June to October, making this one of the few perennial *Laminaria* species worldwide that is known to have a reproductive peak during the summer—this characteristic is usually only found in annual kelps.

HETEROMORPHIC GENERATIONS

Unlike many algal species, day length (short or long days) and temperature do not induce fertility in these sporophytes. Rather, the termination of the active winter and spring vegetative growth period removes the inhibitory compounds produced by the meristem—a zone of rapidly dividing cells at the base of the blade. These compounds had blocked the onset of fertility in the leathery blades of the oarweed. Sporangial sori containing many sporangia surrounded by club-shaped sterile hairs form irregular dark-brown patches on the kelp's surface. Meiosis occurs in the sporangia,

which release zoids during the night. The zoids are typical brown algal motile cells that measure just 3–7 microns in length. They swim exceptionally slowly, settling on the substratum where they grow into gametophytes.

The oarweed has a life history of an alternation of heteromorphic generations in which the large, leathery blades of the sporophyte generation alternate with the microscopic, branched filaments, sometimes consisting of only a few cells, of the gametophyte generation. Gamete formation is induced by a combination of temperatures below 64.5 °F (18 °C), blue wavelengths of light, and the presence of iron ions in the seawater, and results in the formation of one spermatozoid in each cell in the male gametophyte and one egg in each cell in the female gametophyte. Mature eggs are released from the female gametophyte a few minutes after dark. They secrete the chemical sex attractant lamoxirene, which induces the release of the spermatozoids and attracts them to swim to the eggs. The gametes fuse and the zygotes germinate and grow into the large, leathery blades of this species.

→ Large, leathery sporophytes of the oarweed (*Laminaria digitata*) release zoids that germinate and grow into microscopic filamentous gametophytes, usually composed of only a few cells.

KINGDOM	:	Plantae
PHYLUM	:	Rhodophyta
CLASS	:	Bangiophyceae
ORDER	:	Bangiales
GENUS	:	*Porphyra*
SIZE	:	Carpogonium 8–16 microns long
HABITAT	:	Intertidal zone on sheltered and exposed rocky coasts

PHYLUM RHODOPHYTA

Porphyra umbilicalis

Purple laver

Unlike the advanced red algae, some species of the primitive red algae have a life history involving an alternation of either isomorphic or, in the case of purple laver (*Porphyra umbilicalis*), heteromorphic generations.

HETEROMORPHIC GENERATIONS

The reddish-brown thallus of purple laver arises from a well-developed holdfast and fans out into a wide, sheetlike frond that is similar to a fine silk fabric. The thallus, which is only 58–70 microns thick and composed of a single layer of cells, reproduces asexually by neutral spores.

The sheetlike thallus of the purple laver is the gametophyte generation, which alternates with the microscopic sporophyte generation. No specialized reproductive organs are formed by the purple laver. Instead, typical red algal gametes form in the cells on the thallus margins of the haploid male and female gametophytes. One female gamete—the carpogonium—develops in each cell in the thallus margin. It has a small blunt protrusion that differs to the long tubular extension found on the carpogonia of the advanced red algae. Cells of the male gametophytes divide by

mitosis to produce 64–128 haploid male gametes—the spermatia. Propelled by water currents, the spermatium, which lacks flagella and therefore cannot swim, attaches to the surface of and injects its nucleus into the carpogonium. The nuclei fuse to form a diploid zygote. Mitotic divisions of the zygote generate 16 diploid spores, which are released into the sea following the breakdown of the carpogonial wall. Unlike the advanced red algae, which have a life history with three phases, no third phase (the carposporophyte) is formed by the purple laver.

The diploid spores (carpospores) released by the female gametophyte settle on a mollusk shell and bore into the shell's limey matrix. These spores germinate and grow inside the shell into a finely filamentous thallus—the sporophyte generation—that was once known as another species, *Conchocelis rosea*. When mature, the cells in the filaments form diploid conchospores, which divide by meiosis during germination before growing into the sheetlike gametophytes.

NORI PRODUCTION

The discovery of the shell-inhabiting sporophyte of the laver by Dr. Kathleen Drew (see page 242) in 1949 enabled the Japanese to commercialize the production of nori— a close relative of the laver and the delicious seaweed wraps around sushi rolls. The nori industry is now worth US$1.3 billion annually.

→ Large, sheetlike plants of the purple laver (*Porphyra umbilicalis*) are the gametophyte generation that alternates with the microscopic shell-inhabiting sporophyte generation.

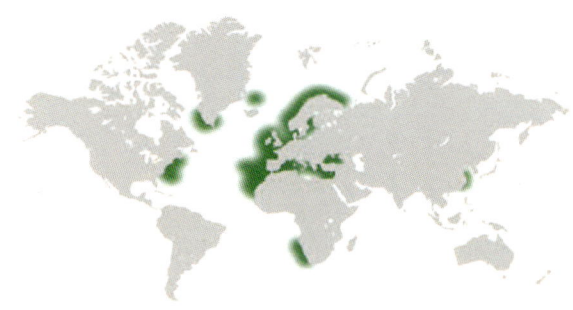

Vertebrata fucoides

Black siphon weed

KINGDOM	Plantae
PHYLUM	Rhodophyta
CLASS	Florideophyceae
ORDER	Ceramiales
GENUS	*Vertebrata*
SIZE	Carpospores 45–66 micron diameter; tetraspores 45–55 micron diameter
HABITAT	Widespread in the intertidal zone

The life history of black siphon weed is an example of the triphasic "*Polysiphonia* type" unique to the advanced red algae, and some phycologists believe it is the most elaborate in all of the macroalgae. It certainly perplexed botanists in the late 1800s, as they observed red algal species with three different types of reproductive cells.

Shigeo Yamanouchi solved the conundrum in 1906, while he was at the University of Chicago in the United States. His revelatory study on the delicate branched filamentous red seaweed *Polysiphonia violacea* (now named *Vertebrata fucoides*) was the first to report a chromosome number of 20 in the gametophytes, gametes, and tetraspores, and 40 in the carpospores and tetrasporophytes of a red alga.

THE *POLYSIPHONIA*-TYPE LIFE HISTORY

The *Polysiphonia*-type life history comprises three phases: the gametophyte, sporophyte, and carposporophyte. Released from the male gametangium on the male gametophyte, the male gamete (spermatium) that lacks flagella drifts passively into the sea until it contacts the long receptive tubular extension on the female gamete (the carpogonium), which

remains attached to the female gametophyte. The spermatium injects its nucleus into the tube; the nucleus travels down the central channel of the tube and fertilizes the female gamete.

Zygotes grow into diploid carposporophytes. These are also retained on the female plant and divide by mitosis to form numerous diploid spherical carpospores that are released into the sea. The carpospores germinate into the diploid sporophytes that, when mature, form tetrasporangia. These tetrasporangia divide by meiosis into four haploid tetraspores, which germinate into the gametophyte generation.

It was thought that the carposporophyte evolved to compensate for inefficiencies in fertilization by spermatia that passively encountered the female gamete. Spermatia cannot swim and red seaweeds lack sex-attracting pheromones—two processes that increase fertilization rates in species that shed their gametes into the sea. A carposporophyte that originates from one zygote and produces numerous carpospores should potentially maintain the population size of red seaweed species.

However, a recent DNA study has shown the opposite: fertilization by spermatia between neighboring plants is, in at least some species, efficient. In another exciting recent study, fertilization rates increased when female gametes of *Gracilaria gracilis* were fertilized by sticky mucilage-embedded spermatia carried by an isopod crustacean species as it moved seeking shelter and food from the male to female plants—the first report in seaweeds of the marine equivalent of pollination.

→ Species of *Polysiphonia* have filamentous thalli and a life history with three phases, typical of the advanced red algae.

ECOLOGY

Almost every habitat on earth

Any mention of seaweeds immediately conjures up images of rocky seashores and beaches. Although seaweeds are the most visible and familiar algae, rocky seashores are just one of a vast range of algal habitats. Amazing adaptations enable different species to live in almost every habitat on Earth, from the equator to the polar regions, from sea level to the mountaintops.

THE UBIQUITOUS ALGAE

In nature, the megadiverse algae are everywhere. Unicellular phytoplankton float or swim in the world's oceans and freshwater environments. Microalgae form mats on mudflats, seaweeds grow on rocky seashores, and pondweeds live in lakes, streams, and ponds. Terrestrial algae inhabit tree trunks, leaves, and fences. Small, temporary puddles, water-filled rock hollows, and birdbaths are often teeming with algae too small to be seen with the unaided eye. There are also some algal species that live in extreme environments such as hot springs, salt lakes, desert soils, and snow. In each of these habitats the algae are free-living organisms, but various species form associations with a diverse array of other organisms. Some algal species may grow on other algal or plant species, or on animals, or they may enter into more intimate relationships as symbionts, or parasites.

Even in seemingly similar habitats, the composition and abundance of species in algal communities may differ markedly in response to subtle differences in the environment. Varying patterns of wave action often determine the seaweed species that grow on rocky shores. On European coasts, the brown seaweed commonly known as thong weed (*Himanthalia elongata*) is abundant on rocks that are exposed to moderate wave action, whereas another brown seaweed, the knotted wrack (*Ascophyllum nodosum*), grows on shores protected from wave action.

Similarly, neighboring ponds will be populated by different phytoplankton species. These variations are the result of an environment selecting those species that are best adapted to the prevailing environmental conditions.

The sheer number of algal species, their widespread occurrence in many habitats, and their differing ecology makes them valuable organisms for the purposes of environmental monitoring and ecosystem health assessments.

→ Two brown seaweeds—the beautifully iridescent-blue rainbow wrack (*Ericaria selaginoides*), framed by thong weed (*Himanthalia elongata*)—dominate this rockpool community on the English south coast.

TERRESTRIAL ALGAE

Algae are not restricted to permanent water bodies. They are also ubiquitous components of terrestrial ecosystems, where they can be found growing on seepages, soils, walls, and mosses, as well as in more aerial habitats on tree trunks and leaves. Most terrestrial algae are cyanobacterial, diatom, and green algal species, with thalli that are either unicells, packets of cells, or short filaments. These algae have developed astonishing adaptations to overcome the rigors of life in adverse terrestrial environments, where the effects of dehydration, fluctuating temperatures, and high solar radiation challenge their survival.

There are approximately 800 terrestrial green algal species, which have thick cell walls wrapped in mucilage as adaptations against desiccation. Microscopic soil algae live on the soil's surface and within soil particles, where they aid in the accumulation of organic matter, prevent erosion, help retain soil moisture, and affect its microbiological activity.

Terrestrial algae frequently colonize the bases of old walls. The humid air near the wall promotes the growth of conspicuous green patches that can form belts stretching several yards in length. Species richness is not high, with 17 green algal and cyanobacterial species found in these patches. In temperate regions, the patches are dominated

← In California, the green alga *Trentepohlia flava* protects itself from the strong sunlight drenching its arboreal habitat, Monterey cypress trunks, by synthesizing its orange sunscreen, beta-carotene.

primarily by species of the green algal genera *Rosenvingiella*, *Klebsormidium*, and *Prasiola*, while *Trentepohlia* and *Cephaleuros* form the green patches on tree trunks in the tropics.

ALGAL ASSOCIATIONS

Various algal species form close associations with other living organisms. They may live on or in other organisms, or may parasitize another species. Some algal species live attached to other algae or seagrasses, or to animals. These commensal associations are generally harmless and common, with the alga merely attaching to a firm surface, as is the case with the diatoms and macroalgae that live on the shell-like backs of turtles. Other algal species live inside the bodies of other algae or animals. These unicellular or small filamentous algae are generally morphologically reduced but have retained their photosynthetic pigments, indicating that they are phototrophs and not parasites. Cyanobacteria live inside sponges; species of a brown algal genus (*Laminariocolax*) live inside kelps; and pennate diatoms reside in thalli of a red alga, (*Coelarthrum opuntia*), and a brown alga (*Ascophyllum nodosum*).

ALGAE AND SYMBIOSES

Algae participate in numerous symbiotic associations. Lichens—partnerships between algae and fungi—are iconic symbiotic associations that are named for the predominant fungus in the associations. Inhabitants of terrestrial environments around the world, they typically grow on trees, smaller plants, fences, rocks and many other surfaces. Around 85 percent of the lichens have unicellular and filamentous green algal symbionts, 10 percent have cyanobacterial partners, and the remaining 4 percent have both green algal and cyanobacterial partners. The common tropical rain forest lichen *Coenogonium linkii* is a partnership between the fungus and the common tropical green alga *Trentepohlia*. In this lichen, four to eight fungal

↖ Numerous microscopic filaments of the tropical lichen *Coenogonium linkii* project horizontally from this tree trunk to construct a thallus that resembles a thin, compact, bracketlike shelf.

↑ An indicator species of moist old-growth forests of cool temperate eastern and western North America, the tree lungwort lichen has declined in comparable European forests due to habitat loss and air pollution.

Symbioses also occur in nutrient-poor tropical and subtropical oceans, intriguingly between various cyanobacterial (*Calothrix rhizosoleniae* and *Richelia intracellularis*) and colonial diatom species (*Hemiaulus, Chaetoceros, Bacteriastrum,* and *Rhizosolenia*). Neither these cyanobacterial nor diatom species flourish in the open oceans outside their symbiotic relationships, which provide the means to exploit an otherwise unavailable oceanic niche. The diatoms have adaptations for buoyancy that enable their cells to stay suspended in the sunlit surface waters. The non-buoyant cyanobacterial species would sink to the ocean floor, away from the light and die if not for the diatoms. The cyanobacterial filaments stay in the sunlit waters by wrapping around the chains of diatom cells (*Calothrix*) or by living inside the diatoms (*Richelia*). The diatoms benefit from the symbioses by receiving the nutrient nitrogen from the cyanobacterial nitrogen fixation, effectively bypassing the nitrogen limitation to their growth in the oceans. These symbioses contribute significantly to the productivity of oceanic food webs.

filaments wrap loosely around the algal filaments, with the spaces among the filaments inhabited by 18 terrestrial diatom species. The lungwort or the lung lichen (*Lobaria pulmonaria*) is a leaflike lichen that is sensitive to air pollution and has disappeared from forests close to major cities and other human developments. It now largely inhabits pristine humid temperate forests in the northern hemisphere. Both a green alga (*Symbiochloris reticulata*) and cyanobacteria (*Nostoc*) are symbionts of the fungus. In lichen symbioses, the alga benefits from the microhabitat constructed by the fungus, which protects it from desiccation and herbivory, "paying" for this protection through the transfer of photosynthetic carbohydrates from the alga to the fungus.

The mass coral bleaching events on coralgal reefs worldwide have showcased the important symbiosis between the reef-building coral polyps and species of the dinoflagellate genus *Symbiodinium* (commonly known by the collective name zooxanthellae; see page 222). Species of *Symbiodinium* also live in the bodies of sponges, sea anemones, jellyfish, giant clams, and nudibranchs, as well as the unicellular ciliates and forams. Zooxanthellae may also be entirely free living, never known to have entered into a symbiotic relationship. Species from eight other dinoflagellate genera have symbiotic relationships with other species, including *Scrippsiella velellae*—the symbiont of the "by-the-wind sailor," *Velella velella,* a beautifully blue jellyfish that floats on the surface of the open oceans.

ALGAL PARASITES

Surprisingly, parasitic species have also evolved in many algal lineages. Approximately 7 percent of dinoflagellates are parasites, infecting crustaceans, annelids, fish, ciliates, diatoms, and other dinoflagellates. These parasites have two different morphologies: a typical free-swimming dinoflagellate cell, and a morphologically different parasitic stage that lives inside a host. These parasitic dinoflagellates may kill their host, which has obvious implications for fish and crustacean aquaculture.

There are also parasitic species among the red, brown, and green algae. Approximately 15 percent of red algal genera occur only as parasites on other red algal species. The majority of these parasitic species are closely related to their host, having evolved directly from that host or from a close relative. Other parasitic species include a brown alga (*Herpodiscus durvillaea*), whose filaments appear as velvety red-brown patches on the surface of the southern bull kelp (*Durvillaea antarctica*); and some species of tropical green algae (*Cephaleuros parasiticus* and *Cephaleuros virescens*) that cause algal leaf spot disease on a wide variety of flowering plants. These green algal parasites have an economic impact when they infect tea, coffee, palm oil, guava, mango, and avocado plants.

↓   The yellow, sausage-shaped parasitic phase of the dinoflagellate *Blastodinium contortum* lives inside the body of crustaceans.

→   The parasitic green alga *Cephaleuros virescens* produces velvety orange-brown patches typical of algal leaf spot disease.

Inland and marine algae

The most diverse and numerous habitats colonized by the algae are found in marine and freshwater environments, although not all aquatic environments on land are freshwater—some are saline. These can range from slightly saline (brackish) to very saline, with a particularly striking example occurring in East Africa, where hot springs deliver alkaline saline water into the rift valley lakes and salt pans.

INLAND SODA LAKES

The phytoplankton community of the East African soda lakes is usually markedly dominated by the spiral filaments of the cyanobacterium *Limnospora fusiformis*, the species used to produce the health food "spirulina" (see page 271). This cyanobacterium attracts dense populations of the lesser flamingo, *Phoeniconaias minor*, rated as one of the most fascinating wildlife spectacles in the world. The lesser flamingo has a highly specialized algal diet, consisting almost entirely of spirulina supplemented with small amounts of pennate diatoms. Blooms of the very fast-growing spirulina support the large flocks. When feeding, the lesser flamingo sweeps its head from side to side just below the water surface, its pistonlike tongue aids in pumping the spirulina-filled water over a filtering apparatus inside its curved beak. This apparatus is equipped with rows of small plates that are the ideal size to collect the spirulina filaments. Alarmingly, massive die-offs of the lesser flamingo have been periodically observed since the mid-1990s; one episode recorded the

↖ A lesser flamingo feeds on the blue-green spirulina bloom in Lake Manyara, Tanzania.

← One million lesser flamingos of Lake Nakuru, Kenya, were estimated to eat the equivalent of 66 tons (60 tonnes) of dried spirulina daily.

deaths of more than 10,000 birds. The causes of these deaths are not fully understood, but they coincided with shifts in the phytoplankton community when a toxic cyanobacterium (*Anabaenopsis*) replaced the spirulina. Along with the toxic effects, the large, slimy, filamentous colonies of *Anabaenopsis* can potentially clog the filtering apparatus of the lesser flamingo, which would also lead to the malnutrition and death of these birds.

ECOSYSTEM ENGINEERS

The example of the flamingos highlights just how important algae are in marine and inland aquatic ecosystems. As primary producers, they underpin aquatic food webs, but some macroalgal species also play a central role as ecosystem engineers, altering their physical surroundings or changing the flow of resources. The most important ecosystem engineers are foundation species that transform the two-dimensional into three-dimensional structures, for example, when individual kelps grow together to form submarine kelp forests arising from the seafloor. In doing so, foundation species create complex habitats that support diverse and productive biological communities. Large brown seaweeds, *Halimeda*, rhodoliths, encrusting coralline red algae, *Spirogyra*, and *Chara* are the foundation species in the communities they engineer.

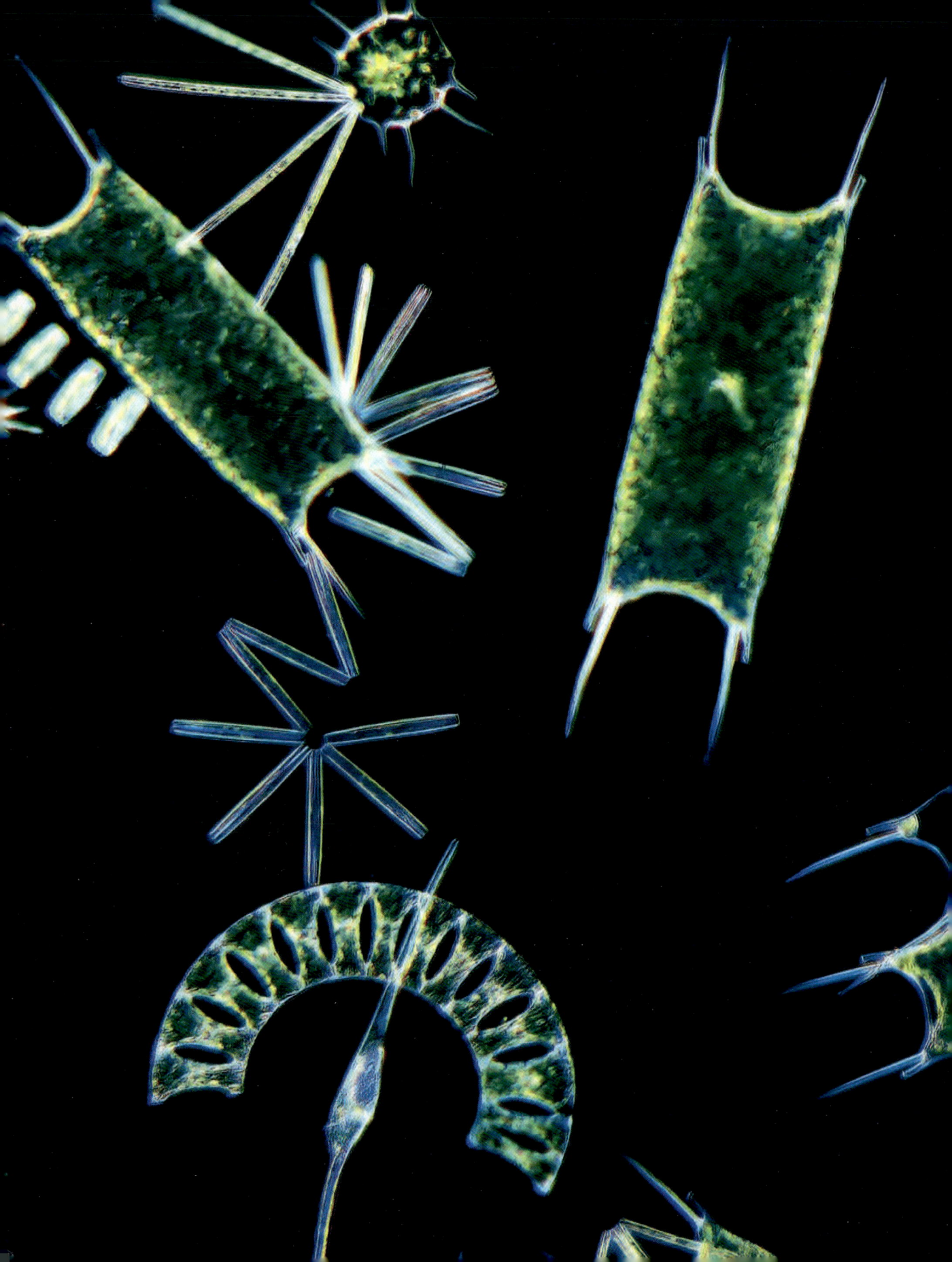

DRIFTERS AND SWIMMERS

The phytoplankton—microscopic unicellular and colonial algal species—drift and swim as the "pastures" in the sunlit waters of the oceans, seas, estuaries, and inland waters. These photosynthetic organisms are important primary producers in aquatic ecosystems. They manufacture the organic compounds that flow through and support aquatic food webs.

MARINE PRIMARY PRODUCTION

Ecosystems vary in their primary production—the rate at which photosynthetic organisms (or primary producers) store energy as organic matter that may flow into the food webs of the ecosystem. Marine primary producers make a significant contribution to global primary production, estimated to be 46 percent compared to the 54 percent contributed by terrestrial primary producers. However, not all areas of the marine environment are equally productive. Rather, the primary production of marine ecosystems is highly variable, and can range from the high levels that often support lucrative commercial fisheries to the low levels characteristic of "marine deserts."

PHYTOPLANKTON CELL SIZE

Phytoplankton are divided into three ecologically important sized classes that are related to their primary production. The largest, the microplankton, vary in size from 20 to 200 microns; the nanoplankton range from 2 to 20 microns; and the smallest, the picoplankton, range from 0.2 to 2 microns.

Despite covering 70 percent of the planet's surface, only one-quarter of marine primary production is generated in the clear blue, low-nutrient "marine deserts" of the open oceans. This habitat is dominated by cyanobacterial picoplankton, whose high surface area to volume ratio of their small cells efficiently strips the low levels of nutrients from the seawater. In addition, many cyanobacterial species can fix atmospheric nitrogen that bypasses the nitrogen limitation typical of oceanic waters.

In many coastal seas, and for most of the year, phytoplankton photosynthesis is limited by the availability of a nutrient (nitrogen, phosphorus, or iron). These nutrient-limited communities are dominated by nanoplankton. However, in coastal areas where water movements transport nutrient-rich water from the seafloor to the sea's surface, the microplankton in the form of larger-celled diatoms have the competitive advantage, storing nutrients in their cells for future growth. In these nutrient-rich areas, the diatoms form natural blooms that terminate once the nutrient supply is exhausted. Diatoms that contribute between 40 and 45 percent to marine primary production predominate in the highly productive areas of coastal seas where they underpin marine food webs, including those that sustain commercial fisheries production.

DRIFTING SEAWEEDS

Macroalgal species that have detached or been forcibly ripped from rocks by storms either sink to the seafloor and decompose, or are stranded on beaches, or they can drift, sometimes for long distances, across the ocean. The drifting thalli remain alive if they can photosynthesize in the sunlit upper layer of the sea. Seaweeds are generally heavier than water and do not float unless they are resuspended in the water column by waves and other water movements, or they have a flotation mechanism.

← Fueled by nutrients accumulated over winter and increased light levels, natural spring phytoplankton blooms, typical of cool temperate coastal seas, are composed of numerous large unicellular and colonial-centric diatoms and fewer colonial pennate diatoms.

Many species of the large brown algae (kelps and fucoids) have pneumatocysts, air-filled bladders that function to buoy the large attached thalli upward to the sea surface. The pneumatocysts also keep drifting thalli afloat while they potentially travel long distances across ocean basins by rafting on the currents. The giant kelp (see page 218) has one large pear-shaped pneumatocyst, 1.5–5 in (4–12 cm) long and 0.4–1.5 in (1–4 cm) in diameter, at the base of each of its many fronds.

Drifting thalli of giant kelp are sometimes encountered in the cool temperate Pacific Ocean. One raft of the giant kelp, estimated to weigh one ton, was encountered by a ship off the coast of Chile, not in sight of land, and in a region of the coast where no attached populations of the giant kelp grew.

The many species of *Sargassum* drift in the sea, buoyed by numerous small pneumatocysts that rarely exceed 1 inch (2.5 cm) in diameter (see page 212). Two *Sargassum* species are particularly well known for forming the large floating *Sargassum* ecosystem in the Sargasso Sea (to which they gave their name) in the middle of the tropical-subtropical North Atlantic Ocean. Another rare adaptation that enables floating is found in the thallus of the southern bull "kelp" (*Durvillaea antarctica* is not a true kelp). The long, leathery blades of this species have a buoyant air-filled honeycomb innermost tissue that is not found in other species of *Durvillaea*.

→ A detached fragment of the giant kelp *Macrocystis pyrifera* floats off the Californian coast, buoyed by an oval pneumatocyst at the base of each leaflike frond.

Freshwater algal ecosystems

The green algae and the diatoms are the most species-rich and common freshwater algae. Unicellular and colonial green algal species (phylum Chlorophyta) and the diatoms are important components of phytoplankton communities that contribute significantly to food webs in lakes, streams, ponds, and rivers. Many species are indicators of ecosystem health, which is important for environmental management. For example, the green alga *Pediastrum boryanum* is an indicator of nutrient-rich water.

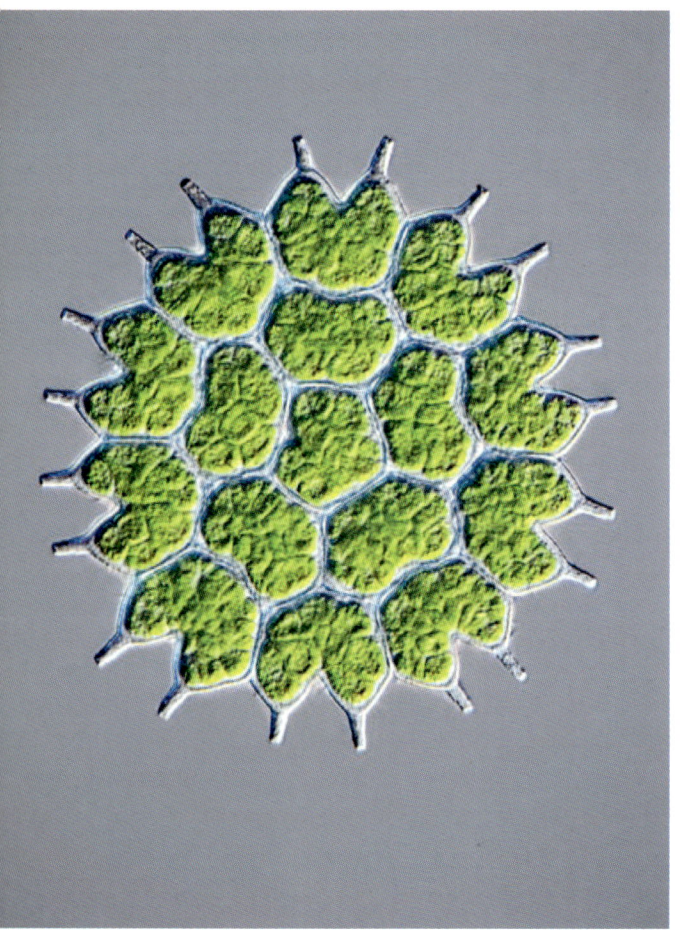

MATS

The other green algae—the charophytes—are important ecological dominants in certain freshwater habitats. Unicellular charophytes—the desmids—that are numerous and diverse in peat bogs and ephemeral freshwater pools are valuable indicators of ecosystem health in these habitats.

Filamentous charophytes, particularly *Spirogyra* and *Zygnema*, dominate various freshwater habitats and quickly produce extensive mats. *Spirogyra* mats of more than 18 in (45 cm) thick have washed up on the shores of Lake Huron in the North American Great Lakes. Inhabiting lakes, ponds, ditches, and slow-flowing streams around the world, *Spirogyra* can grow as mats or streamers, with the latter reminiscent of long tresses billowing out in the current.

Zygnema has markedly different habitat preferences, forming extensive mats in wet terrestrial environments in the Arctic and Antarctic, where it is well adapted to the harsh environment. Species of the two genera are

↗ Trapped oxygen bubbles float the luxuriant mats of *Spirogyra* at the well-lit water surface, maximizing the rate of photosynthesis and mat growth.

← A microscopic colony of *Pediastrum boryanum*, a species often found in the freshwater phytoplankton.

among the main primary producers and influencers of nutrient cycling in their habitats, with *Zygnema* mats also playing an important role in the primary colonization and development of soils.

UNDERWATER MEADOWS

Standing erect and attaining heights of approximately 3 ft (1 m), the thalli of *Chara* and *Nitella* can form dense underwater meadows of considerable biomass in alkaline, calcium-rich freshwater lakes, and in streams with low concentrations of nutrients.

These algal communities are an important food source for waterfowl and create a nursery area for fish. They also prevent sediment resuspension, which reduces water turbidity and stabilizes the clear water state of shallow lakes. *Chara* and *Nitella* are common in shallow to deep freshwater lakes and are generally thought to be adapted to low-light habitats, but these populations can decline and eventually disappear when increasing nutrient enrichment leads to eutrophication.

Mudflat algae

Unlike the land plants whose roots anchor the plant into the soil, most macroalgae are unable to grow on the shifting soft bottoms of sand and mudflats. The rhizoids of the macroalgae secrete an adhesive that can only attach the thallus onto a firm, often rock, surface. In nature, however, there are always exceptions. A few macroalgal species have adapted to either growing on mud or on the aerial roots of mangroves.

RUNNERS AND BULBOUS HOLDFASTS

One group of tropical green seaweeds can form extensive meadows on soft bottoms. Their runners and bulbous holdfasts adapt them to life in these habitats. Thalli are anchored into the soft bottoms in species of *Caulerpa* by horizontal stemlike stolons that run across the sand, and produce, at regular intervals, bundles of rootlike rhizoids, and in *Udotea* and *Halimeda* by a taproot-like bulbous holdfast of unorganized siphons.

↗ This light micrograph of the red seaweed *Caloglossa leprieurii* reveals the pattern of cells on two of many elegant membranous blades that grow as a rosette on the trunks and aerial roots of mangrove trees.

← Thick turfs of the *Bostrychia–Caloglossa* association growing on the peglike aerial roots of the gray mangrove are exposed at low tide to the harmful effects of the dry atmosphere.

MICROSCOPIC MATS AND BIOFILMS

Although appearing to lack any algal life,
the muddy areas on mudflats are covered by the
microphytobenthos—microscopic algal communities
composed of either benthic mats of filamentous
cyanobacteria and pennate diatoms, or biofilms of
pennate diatoms. These important communities, which
are responsible for the majority of primary production
on mudflats and provide food for deposit-feeding
animals, occupy the upper millimeters of the mud,
giving the mud surface a subtle brownish or greenish
tinge. Within this community, the cyanobacteria and
diatoms secrete a layer of extracelllular carbohydrate
onto the sediment surface to glide across while
photosynthesizing at low tide. This also has the benefit
of improving water quality, as the sticky carbohydrate
binds and stabilizes the sediments, decreasing the
resuspension of mud into the water column.

MANGROVE ALGAE

Species of the red algal genera *Bostrychia* and *Caloglossa*
are widely distributed in marine, estuarine, and
freshwater habitats from temperate to tropical regions
worldwide. These genera are also the community
dominants in the *Bostrychia*–*Caloglossa* association, the
red algal-dominated community that inhabits the
trunks and aerial roots of mangrove trees. Mangrove

trees are the mudflat specialists of tropical and
subtropical coasts, where they grow on waterlogged
airless soils. Airless soils limit the mangrove's ecological
distribution to the upper intertidal zone, where
their above-ground aerial roots are exposed to the
atmosphere for three or more hours during low tide.
This long exposure provides sufficient time for the
diffusion of oxygen from the atmosphere through the
aerial roots and into the mangrove's subterranean root
system. Mangrove roots require oxygen for respiration.
The mangrove algae have developed adaptations to
survive the harsh environment of tropical upper
intertidal zones. Species of *Bostrychia* and *Caloglossa*
tolerate the increasing salinity during long periods of
aerial exposure by synthesizing and accumulating large
organic molecules; in the case of *Caloglossa*, this is the
sugar alcohol mannitol. The macroalgae also grow as
a thick turf on the aerial roots, which not only traps
water among the thalli, but also reduces the surface
area from which water can be lost.

Coralgal reefs

Coralgal reefs occur in tropical and subtropical seas, and are important ecosystems, supporting as many as 1.3 million species and more than 500 million people worldwide. They are structured by a complex interaction between the coral polyps and the algae, prompting phycologists to rename coral reefs, "coralgal reefs."

ALGAL STRUCTURED REEFS

The importance of algae on coralgal reefs is not limited to the coral polyp's symbiotic algae, the zooxanthellae. The algae directly or indirectly contribute to constructing the reef, as well as determining the structure of the reef communities. This is evident from their ability to potentially change the character of the reef in response to natural or human-mediated disturbances that alter the reef's nutrient regimes and grazing patterns.

THE TROPHIC PYRAMID

The important role of algae on coralgal reefs is reinforced by one of the basic tenets of biology—the trophic pyramid of biomass. In the "traditional" trophic pyramid, the biomass (weight) of the primary producers should be approximately 10 times the biomass of herbivores, and 100 times that of the first order carnivores. The coral polyp is not a primary producer, instead deriving a large proportion of its energy from the photosynthesis of its zooxanthellae. The algae are the primary producers of healthy coralgal reefs, and they conform to the trophic pyramid. This fact was recognized as early as 1955 when the Eniwetok reef flat in the Pacific Ocean was reported to be dominated by algae, which accounted for 85.1 percent of the biomass, compared to 13.8 percent for the herbivores and 1.1 percent for the carnivores. Most striking was the finding that the zooxanthellae constituted only 20 percent of the algal biomass, with macroalgae making up a further 6 percent, and the algal turfs, endolithic algae, and algal crusts a whopping 73 percent. This has been substantiated in the last 60 years of research.

Coralgal reefs are best developed in tropical seas, in the path of the trade winds, where both local wave-generated and equatorial currents drive huge volumes of low-nutrient seawater across the reef. The key is the huge volume, as it fuels the algal growth.

There are four different thallus forms that commonly occur on coral reefs: turf algae, encrusting coralline algae, coarsely branched macroalgae, and unicellular zooxanthellae. These thallus forms attain their greatest biomass in different habitats on coralgal reefs. Encrusting coralline algae dominate the reef's seaward edge; the turf algae and coral polyps with their zooxanthellae inhabit the often extensive outer reef flat behind the seaweed edge, and the coarsely branched macroalgae dominate the inner reef flat and lagoon on the landward side of the reef.

→ The green seaweed *Anadyomene menziesii* (top right), the brown seaweed *Dictyota* sp. (center left), and crusts of the red seaweed *Peyssonnelia* (center right) grow with corals on Pulley Ridge coralgal reef, Gulf of Mexico.

TURF ALGAE

The turf algal communities of coralgal reefs are master primary producers in nutrient-poor seas. Composed of assemblages of many species of cyanobacteria, as well as red, green, and brown algae, the filamentous turf algae with their high surface area to volume ratio are highly efficient at stripping the nutrients nitrogen and phosphorus out of low-nutrient seawater. This was recognized four decades ago when large public aquaria in the tropics started using turf algae to reduce the high nutrient loads delivered into aquarium seawater in the excreta of the resident fish, sharks, rays, and turtles. During the daylight hours, the aquarium seawater is pumped continuously onto the sunlit roof of the aquarium, where it runs over turf algae grown on a series of corrugated panels. Resembling the washboards of old, the panels are placed in troughs to catch the seawater, which is then returned to the aquarium.

Turf algae are far more efficient at stripping nutrients from seawater than any known chemical method, and it is also a cheaper, natural, and nonpolluting method for the aquarium. Another benefit is that the excessive algal growth is regularly scraped off the panels to maintain the high growth rates, providing the resident herbivorous green turtles in the aquarium with a meal of delicious turf algae.

Cyanobacteria, which have the highest biomass within the algal turf communities, play another important role on healthy reefs. They fix nitrogen, adding to the nitrogen budget of the reef and so overcoming the nitrogen limitation that commonly limits algal growth in the marine environment.

On coralgal reefs, the turf algae are generally less than 1.2 in (3 cm) tall, have slender branches to 0.4 in (1 mm) in thickness, and often have a creeping habit. *Herposiphonia secunda* has a morphology typical of many

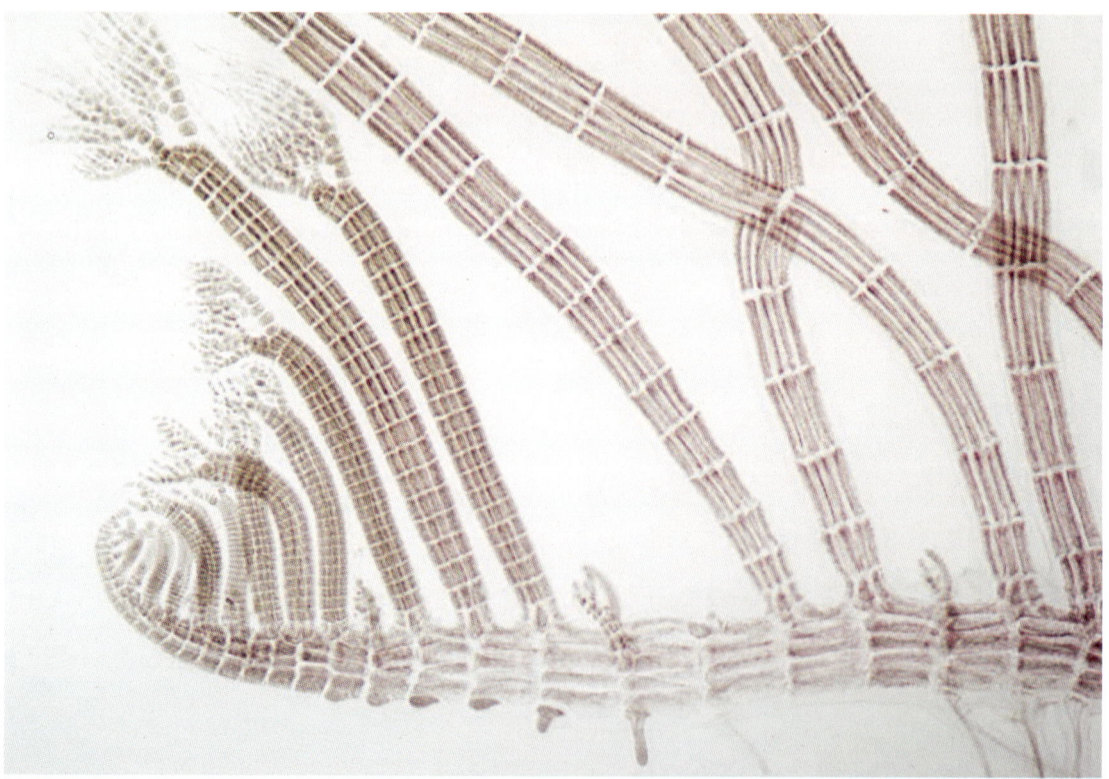

← Species of the branched red seaweed *Chondria* (seen here growing with small purple and green invertebrates) are common in algal turfs on coralgal reefs.

↙ Adapted to the intense herbivory on coralgal reefs, *Herposiphonia secunda* grows from the upcurled tip (bottom left) of its creeping thallus. The survival of the thallus is unaffected when its erect branches are eaten by herbivores.

turf algae in that its branched horizontal axes that attach to the substratum at numerous points give rise to erect branch systems.

Turf algae dominate on healthy coralgal reefs, growing in crevices and on lateral rock surfaces, and covering more than 50 percent of the substratum that is not covered by living coral polyps. In other marine ecosystems, turf algae superficially resemble the grass turfs that grow on land. However, turf algae are inconspicuous components on healthy coralgal reefs, with most of their thalli consumed by grazing fish (parrotfish, surgeonfish, and damselfish) and invertebrates (sea urchins, marine snails, crabs, and crustaceans). Many turf algal species have evolved, as adaptations to intense grazing, creeping axes that cling to the substratum in small crevices and cavities in the reef matrix away from the reach of the herbivores. The creeping axes have a high regenerative capacity and continue to produce more erect branches almost at the same rate as they are consumed by the herbivores. This delicate balance between the growth of the turf algae and herbivore grazing is disrupted by human activities on coralgal reefs. The delivery of increased nutrient levels, which fuels excessive algal growth or overfishing, which decreases herbivore grazing, leads to the overgrowth of the coral by the turf algae.

THE CORAL SYMBIONT

For its nutrition, the coral polyp depends on its photosynthetic symbiont, the zooxanthellae, which are dinoflagellates of the genus *Symbiodinium* (see page 222). Organic molecules produced by *Symbiodinium* during photosynthesis are transferred to the coral polyp in return for the polyp's useful waste products, including ammonia and nitrate (which are typically limited on reefs), and carbon dioxide. In addition, the calcification of reef-building corals appears tightly linked to photosynthesis of their zooxanthellae, although the processes remain unclear. Nevertheless, the zooxanthellae enable the coral polyps to form calcium carbonate skeletons that build some of the largest living structures on Earth.

← Red algal coralline crusts and small tufts of the green seaweed *Microdictyon umbilicatum* (lower third of the image) grow with corals on pristine Swains Island reef, American Samoa, South Pacific Ocean.

↙ Lower image. On the seaward reef edge, *Porolithon onkodes* grows as extensive rock-hard pavements that protect the reef from destructive oceanic waves.

↘ The eye-catching bright-green turtle weed (*Chlorodesmis fastigiata*) of Indo-Pacific coralgal reefs contains noxious chemicals that deter fish grazing.

REEF BUILDERS

Certain algal species are essential components of the carbonate rock (limestone) on which present-day coralgal reefs are built. Cores drilled through reefs have identified the calcareous organisms whose skeletons have built this carbonate rock. On open ocean atolls, the biotic sources of this rock—in order of volume—are encrusting coralline red algae, forams (unicellular calcareous animals), the green alga *Halimeda*, and reef-building corals.

Encrusting coralline red algae are the ecosystem engineers that build, maintain, and protect coralgal reefs. They dominate the shallow, turbulent areas of the seaward reef edge that is impacted by the full force of high-energy oceanic waves. Here, their spreading growth habit forms a hard cement over the reef surface that, over centuries, has built up a massive, slow-growing pavement. This seaward reef slope, which extends from a depth of 65 ft (20 m) to just below low tide mark, protects the delicate coral polyps and other algae on the reef flat from the persistent destructive oceanic waves. The encrusting coralline red alga *Porolithon onkodes* has been regarded as the most common and important algal ridge builder worldwide. However, recent DNA sequencing studies have suggested that there may be as many as 20 species currently recognized as this single species.

INNER REEF FLAT AND LAGOON

Sitting above the reef slope around low water mark, the reef
rock ridge is responsible for damming seawater to depths of 12 in
(30 cm) across the reef flat and lagoon at low tide. Coarsely
branched macroalgae are most common and species-rich in the
pool-like inner reef flat where they grow on the rocks, rubble, or
dead coral. The most conspicuous inhabitants of this zone are the
large tropical brown algae—*Sargassum*, *Hormophysa*, and *Turbinaria*—
that form a canopy over an understory of smaller red, green, and
brown macroalgae. Siphonous green algae—particularly *Halimeda*
and *Caulerpa*—and seagrasses (flowering plants that colonized the
sea from the land) are commonly found growing on the sandy
bottom of the lagoon.

↑   Sea grapes (*Caulerpa racemosa*)
and *Halimeda micronesica* are common
inhabitants of coralgal reefs in the
Indo-West Pacific, growing to depths
of at least 121 ft (37 m).

Rocky shore macroalgae

Dense populations of macroalgae that exhibit distinct patterns of zonation frequently dominate rocky seashores. These patterns—and the mechanisms that determine them—have fascinated phycologists for more than a century. Among the important physical factors that determine the ecological distribution of these macroalgal species are the tides above low tide mark and light penetration below low tide mark.

↑ Tough calcified coral weed (*Corallina officinalis*) withstands desiccation at low tide, strong wave action, and grazing.

→ A large, pluglike holdfast of the southern bull kelp *Durvillaea antarctica* anchors this surf zone specialist to rocks on rough-water New Zealand and Chilean coasts.

INTERTIDAL SEAWEEDS

Macroalgae growing in the intertidal zone (the area between low tide and high tide marks) are exposed to the atmosphere at low tide. How much exposure the algae receive will vary, depending on the height of the shore and the height of the tide. Macroalgae growing on the upper shore may be exposed at low tide for two or more hours, whereas the southern bull kelp (*Durvillaea antarctica*) and other seaweeds growing around low tide mark may be exposed for less than one hour, for a couple of days at the two-weekly spring tides (the highest high tides and the lowest low tides of the month).

The area between the low and high tide marks is divided into high, mid, and lower intertidal zones, based on exposure time to the atmosphere. Macroalgal species that inhabit the intertidal zone usually grow in one of the three zones. Coral weed (*Corallina officinalis*) grows most abundantly in the mid intertidal zone. Low tide exposes intertidal macroalgae to a variety of potentially adverse environmental factors, including desiccation, extreme temperatures (hot and cold), high light levels, and osmotic shock. In the more stable subtidal environments there is no risk of desiccation and therefore osmotic shock; the light levels are lower, as the water absorbs the light passing through it, and the temperature is favorable and shows little variation.

UPPER SHORE SEAWEEDS

The upper distribution limits of macroalgal species on the seashore are set by physical factors—many species simply do not survive the temperature and desiccation stress of being high and dry for the prolonged periods that characterize the upper intertidal zone. High intertidal specialists, the extremely dessiccation-tolerant *Porphyra linearis*, can survive three weeks without being covered by the tide, and the channelled wrack (*Pelvetia canaliculata*) four to six days, but this period reduces to around two hours in bladder wrack (*Fucus vesiculosus*)

and knotted wrack (*Ascophyllum nodosum*), which live lower on the seashore. Other macroalgae employ a variety of strategies to overcome living in the intertidal zone: the saccate seaweeds *Colpomenia* and *Halosaccion* avoid desiccation stress by drawing on their central sacs of stored seawater.

LOWER SHORE SEAWEEDS

The lower distributional limits for macroalgal species on intertidal rocky shores can also be set by physical factors on wave-exposed coasts, but on semi-exposed

and sheltered shores they may be determined by biotic interactions. Space for the attachment of macroalgae on rocky shores is often limited, leading to competition among the species. This, together with the pressures of herbivory, which consumes either or both the juvenile or adult seaweed, determines the structure of lower intertidal communities. Below low tide mark, space, herbivory, and light levels determine the structure of subtidal macroalgal communities.

Exposure to wave action often determines the composition of macroalgal communities. Some species flourish on wave-swept coasts. The sea palm *Postelsia palmaeformis* (see page 216) grows only on rough water rocky shores in the northeast Pacific. The southern bull kelp (*Durvillaea antarctica*, see page 191) inhabits the upper subtidal zone of some wave-swept coasts in the cool temperate southern hemisphere. The large, pluglike holdfast firmly attaches the large, leathery thallus to the rocks and the thick stipe and thick leathery streamlined blades are adapted to the drag forces imposed on the thallus by the breaking waves. The wave-swept articulated coralline alga *Calliarthron cheilosporioides* is also a surf zone specialist. Its erect thalli are constructed of alternating calcified and narrow uncalcified segments; the latter function as flexible joints. The stony calcified segments protect the plants from the physical damage of the crashing waves, while the strong flexible joints enable the thallus to bend, which prevents dislodgement from the rocks under the force of the crashing waves.

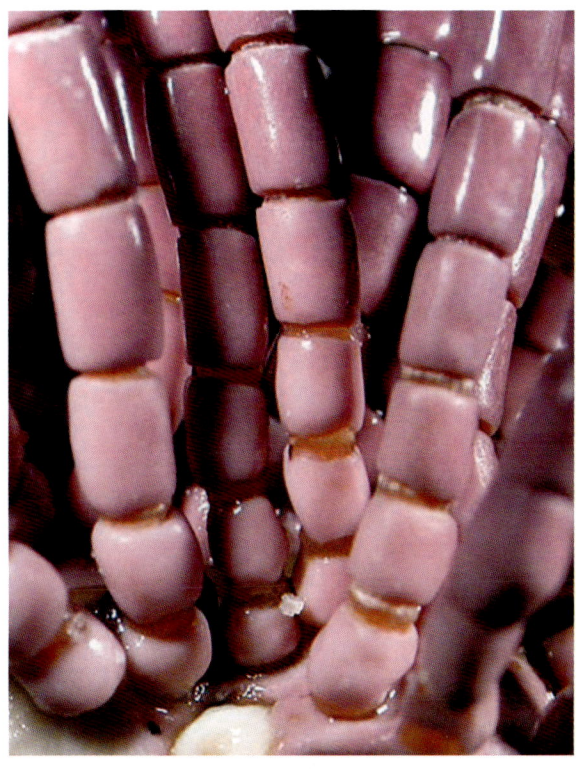

↑ A surf zone specialist, *Calliarthron cheilosporoides* thrives, often growing in wide bands, on wave-swept rocky shores in central California. Its hard but flexible thallus survives a life lived in crashing waves.

← One of the most desiccation-tolerant seaweeds, channelled wrack (*Pelvetia canaliculata*) inhabits the upper intertidal zone on sheltered and wave-exposed European coasts from Iceland to Portugal.

Kelp and fucoid forests

Kelps and fucoids are two groups (orders) of large brown seaweeds that form conspicuous and ecologically important ecosystems, either worldwide (fucoids) or from temperate to polar coasts (kelps). Some of these species grow in the intertidal zone. Others are ecosystem engineers that construct submarine forests where they are the key primary producers, ameliorating the environment to provide shelter and complex habitats for other algae and diverse assemblages of invertebrates and fish.

GIANT KELPS

The most spectacular giant kelps—bull kelp (*Nereocystis luetkeana*), elk kelp (*Pelagophycus porra)*, and giant kelp (*Macrocystis pyrifera,* see page 218)—have by far the largest algal thalli, growing to more than 100 ft (30 m) in length. Bull kelp (see pages 116 and 208–9) and elk kelp are endemic to the northeast Pacific coast, and while giant kelp is restricted to this coastline in the northern hemisphere, it is also widely distributed on cool temperate coasts in the southern hemisphere. As the center of kelp diversity, the northern Pacific coasts are considered the sites for kelp evolution, with giant kelp thought to have crossed the equator to colonize southern hemisphere shores some time between 10,000 and 3.1 million years ago.

Smaller kelp species differ in their geographical distribution. *Laminaria*, the genus with the most kelp species, is widely distributed on coastlines in both hemispheres. *Lessonia* is a southern hemisphere genus, occurring solely in South America, New Zealand, and southern Tasmania, Australia. *Ecklonia* is the most warm-water tolerant kelp genus, common along warm and cool temperate coasts in the Indian and Pacific Oceans.

The order Laminariales is poorly represented in southern Australia, which records only three kelp species: the giant *Macrocystis pyrifera*, plus *Ecklonia radiata*, and *Lessonia corrugata*. However, this coastline has a high species richness overall, including numerous endemic species of the order Fucales.

← Elk kelp (*Pelagophycus porra*), with an upper thallus of distinctive antlerlike branches, has a narrow geographical range from the Channel Islands of the southern California coast to the central Baja California, Mexico. It is ecologically restricted to cool, deep (66–164 ft/20–50 m) habitats, seaward of the *Macrocystis* forests.

FUCOID FORESTS

Fucoids dominate brown algal communities in the intertidal zone on European coasts and some temperate coasts elsewhere in the world. On British coasts, three species of *Fucus* form well-defined bands in the intertidal zone: *Fucus spiralis* in the upper intertidal, *Fucus vesiculosus* in the mid intertidal, and *Fucus serratus* in the lower intertidal zone. In comparison, just one intertidal fucoid, *Hormosira banksii* (commonly known as Neptune's necklace), forms an extensive band in the mid and lower intertidal zones on the temperate coasts of southern Australia and New Zealand.

Large brown algal species of the order Fucales also form submarine forests. However, they rarely grow more than 15 ft (around 5 m) in length and they never attain the size of the giant kelps. The fucoids are

nevertheless still ecosystem engineers, constructing communities in which they are the canopy species, providing a substrate, food, and habitat for diverse assemblages of other algae, invertebrates, and fish, as well as spawning and nursery grounds for invertebrate larvae. Many understory macroalgal species live on the rocky substratum below the canopies in these submarine forests, ranging from dense velvety turfs to frondose thalli to coralline red algae.

On European coasts, thong weed (*Himanthalia elongata*) and the knotted wrack (*Ascophyllum nodosum*) are common habitat-forming fucoid species. Species of *Cystoseira* and their close relatives are the key foundation species in the fucoid-dominated communities of the warm temperate eastern Atlantic Ocean, including the offshore islands of the

Macaronesian archipelago and the Mediterranean Sea, which has the highest species richness of these large brown algae in the world. *Cystoseira* forms extensive forests on rocky substrata on both sheltered and wave-exposed shore, ranging from intertidal rock pools to depths of 165 ft (50 m), and supports diverse communities.

←   The fucoids thong weed (*Himanthalia elongae*) and toothed wrack (*Fucus serratus*), and oarweed kelp dominate this lower intertidal shore at Cornwall, England.

↓   Dense strings of buoyant, beadlike vesicles equip Neptune's necklace (*Hormosira banksii*) to the rigors of life on temperate Australian and New Zealand intertidal shores.

TROPICAL FUCOIDS

While many large brown algal species, particularly the kelps, are restricted in their geographical distribution to polar and temperate coasts, four fucoid genera grow in tropical seas. Three are restricted to the coasts of the tropical Indian and west Pacific Oceans and some islands of the South Pacific Ocean, whereas *Sargassum* is abundant and geographically widespread from tropical to cool temperate seas. An ecologically successfully genus of 359 currently accepted species, it is highly likely that many more *Sargassum* species will be discovered when poorly known tropical ecosystems are fully investigated. Species of *Sargassum* are the ecosystem engineers of submarine communities on tropical to warm temperate coasts worldwide but their maximum thallus height of 6.6 ft (2 m) ensures that these communities never attain the complexity of the giant kelp forests. On coralgal reefs, where they inhabit the reef lagoons, species of *Sargassum* typically form large beds, providing the structure for a major reef habitat, as well as increasing reef biodiversity.

Deep–sea macroalgal communities

Light is quickly absorbed as it passes down to the depths of the ocean. Only the green wavelengths in cloudy coastal waters and blue-green wavelengths in clear oceanic waters reach the lower limit of phototrophic life. Light is generally considered the controlling factor that limits macroalgal photosynthesis and growth with increasing depth. Some deep-sea macroalgal communities are floristically unique, with species restricted in their ecological distribution to these low-light environments.

LIGHT LEVELS

Macroalgae that live below the water surface are well adapted to the low-light regimes typical of these environments. They employ many strategies that assist in absorbing the available light. Generally, leathery macroalgae, including kelps, can grow down to depths where only 1 percent of the sunlight that strikes the ocean surface is available for photosynthesis, delicate leaflike macroalgae 0.1 percent, and encrusting coralline red algae about 0.01 percent. Red seaweeds such as *Halymenia* sp. and *Kallymenia westii* are often more common than green or brown seaweeds in deeper waters. The red photosynthetic pigment phycoerythrin is extremely efficient at scavenging the green wavelengths of light that are then transferred to chlorophyll to be used in photosynthesis.

WORLDWIDE DISTRIBUTION

Deep-sea macroalgal communities can be found thriving around the world: in the Mediterranean Sea, fucoid communities grow at depths of around 80–165 ft (25–50 m); on the central Californian coast, kelps dominate at depths from around 100–150 ft (approximately 30–45 m); red algae grow at depths of 130–180 ft (40–55 m); and encrusting coralline red algae form infrequent patches at depths of 180–245 ft (55–75 m). Many species of siphonous green algae— especially the genera *Udotea*, *Penicillus*, and *Halimeda* (see page 214)—also form deep-sea communities. In the Hawaiian Islands, *Udotea geppiorum* creates meadows at depths of 65–295 ft (20–90 m), peaking at mesophotic ("middle light") depths of 195–280 ft

(60–85 m), while meadows of *Penicillus capitatus* grow at depths of 90–165 ft (28–50 m) in the Canary Islands. In Australia's northern Great Barrier Reef lagoon, *Halimeda* forms extensive communities at depths of 65–130 ft (20–40 m).

Species of the sea lettuce (*Ulva*) are among the deepest growing green seaweeds, recorded at depths of 655 ft (200 m) in the Hawaiian Islands, but encrusting coralline red algae have been found at even greater depths—these red algae have been observed growing at a depth of almost 879 ft (268 m) off an uncharted seamount near the Bahamas in the tropical western Atlantic Ocean. In this extreme low-light environment, less than 0.001 percent of the light falling on the sea surface is available for photosynthesis.

← The red seaweeds *Halymenia* (center) and *Gracilaria blodgettii* (right) grow at depths of 225 ft (75 m) in the Gulf of Mexico.

→ Shaving brush alga (*Penicillus capitatus*) is a subtidal tropical to warm temperate green seaweed.

Extremophiles

It is unsurprising that some species of the megadiverse algae have an enormous adaptive capacity that enables them to survive in some of the most extreme environments on Earth. A variety of algal species have been found in salt lakes, snow, ice, hot springs, deserts, soil crusts, acidic lakes, and other extreme environments.

SALT LAKE ALGAE

Hypersaline environments—those with a salinity greater than 5 percent compared to the approximately 3.5 percent for seawater—select for halophytes, species adapted to high salt concentrations. Many algal species that lack the physiological adaptations to cope with the toxic salt are excluded from these environments. Common inhabitants of hypersaline environments are species of the Chlorophyta (particularly *Dunaliella*), cyanobacteria, and diatoms.

Many of the 28 species of the green algal genus *Dunaliella* are broadly tolerant halophytes, increasing the concentration of the large organic molecules of glycerol inside their cells to counteract the toxic osmotic effect induced by the hypersaline environment. As its name implies, the planktonic flagellate *Dunaliella salina* is one of the most salt-tolerant eukaryotes, capable of surviving concentrations greater than 10 percent salt, while *Dunaliella acidophilia* is an acid-loving species that can survive in extremely acidic environments with a pH as low as 2.

SOIL ALGAE

Soil crusts play important ecological roles in primary production, nitrogen fixation, water retention, and the stabilization of soils in extreme arid, semiarid, and alpine environments. The algae that inhabit soil crusts have to contend with desiccation, extreme temperature fluctuations, and high solar radiation. Certain green algal and cyanobacterial species have achieved ecological success in these environments by developing adaptations that limit water loss, including thick cell walls, mucilage layers, aggregating cells into packets, and the production and accumulation of "water holding" organic molecules. The green alga *Neochlorosarcina* survives in its soil habitat utilizing these water conserving adaptations. Its thick walled cells are organized into three-dimensional packets (sarcina) and they secrete large quantities of mucilage that completely envelope them.

RED HOT ALGAE

Some species not only survive in extreme conditions, they positively thrive in them. For example, unicellular red algal species of *Cyanidium*, *Cyanidioschyzon*, and *Galdieria* inhabit thermoacidic environments with a low pH of 0.2–4.0 and high temperatures of 107–135 °F (42–57 °C). Found in hot springs, soils, seeps, and the pore space of rocks in geothermal areas worldwide, their sheer numbers and brilliant blue-green color visually identifies them as the dominant organisms in these extremely hostile habitats.

→ A light microscope image of packets of cells of the green alga *Neochlorosarcina* growing on soil in an ancient Scottish woodland. The background stained with black ink highlights the whitish mucilage surrounding the bright green cells.

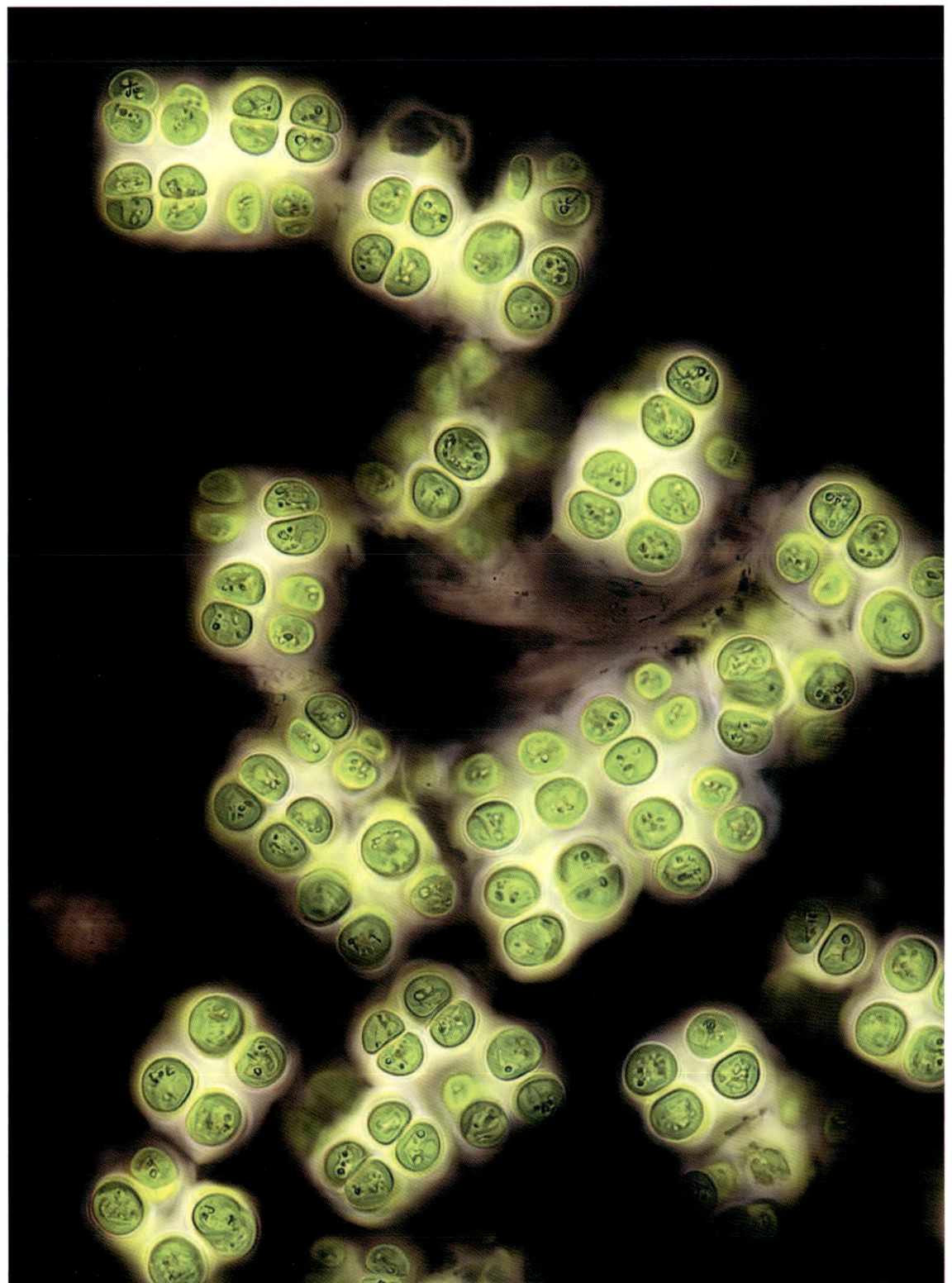

Algal predators

Species of euglenoids and dinoflagellates display diverse nutritional strategies that include phototrophy, heterotrophy, and mixotrophy. Both groups have heterotrophic species that lack plastids, and these species feed by absorbing dissolved organic matter, eating bacteria, or ingesting prey. Species of euglenoids that ingest prey have evolved an elaborate feeding structure, and the helical strips of their cell wall slide against each other, enabling the cell to change shape while consuming its prey.

MIXOTROPHS

Colorless dinoflagellates that ingest prey have long been known, but it was a surprise to find that some dinoflagellates with plastids are also predators, and therefore mixotrophs. Dinoflagellate predators feed on other phytoplankton cells, ciliates, crustacean (copepod) eggs, and the larval stages of invertebrates; in some instances they will immobilize their motile prey by trapping them with capture filaments, or paralyzing them with a toxic mucus secretion.

The cell covering—with (armored species) and without plates (unarmored species)—determines how the prey is taken into the dinoflagellate cell. Without plates acting as a barrier to ingestion, unarmored species engulf and take the intact prey into the cell. Some unarmored and armored species insert a feeding tube (the peduncle) and suck the prey's protoplast into the dinoflagellate cell. In the third feeding behavior, armored dinoflagellates use a veillike feeding extension of the cytoplasm (a pallium) that emerges from the cell and envelopes the prey. A few armored species can also ingest their prey whole, using a large opening in the longitudinal groove (sulcus) between the plates.

A PREDATOR

The more than 250 colorless armored species of the genus *Protoperidinium* have cells of different sizes and morphology. Cells of the various species are small to large (16–250 microns long) and are either spherical or have spectacular horns. Some species in the genus are known to use a feeding veil. Named for its voracious habit, *Protoperidinium vorax* is a small (16–26 microns long) spherical species that feeds on unicellular and chain-forming colonial centric diatoms (*Thalassiosira* spp., *Skeletonema costatum*, *Chaetoceros* spp., and *Leptocylindrus danicus*). Whenever *Protoperidinium vorax* hits a prey cell, it stops swimming, rotates, and aligns its longitudinal groove toward the prey. After a few seconds, the feeding veil emerges from the groove at the point of contact between the dinoflagellate and the diatom, and surrounds the diatom cell. The diatom's protoplast changes consistency and is sucked into the veil; when feeding is complete, the veil is withdrawn into the groove, leaving the empty diatom cell.

→   A light microscope image of the predatory unicellular dinoflagellate *Protoperidinium*, which has extended a feeding veil from its cell to envelope its prey, a colony of four diatom cells.

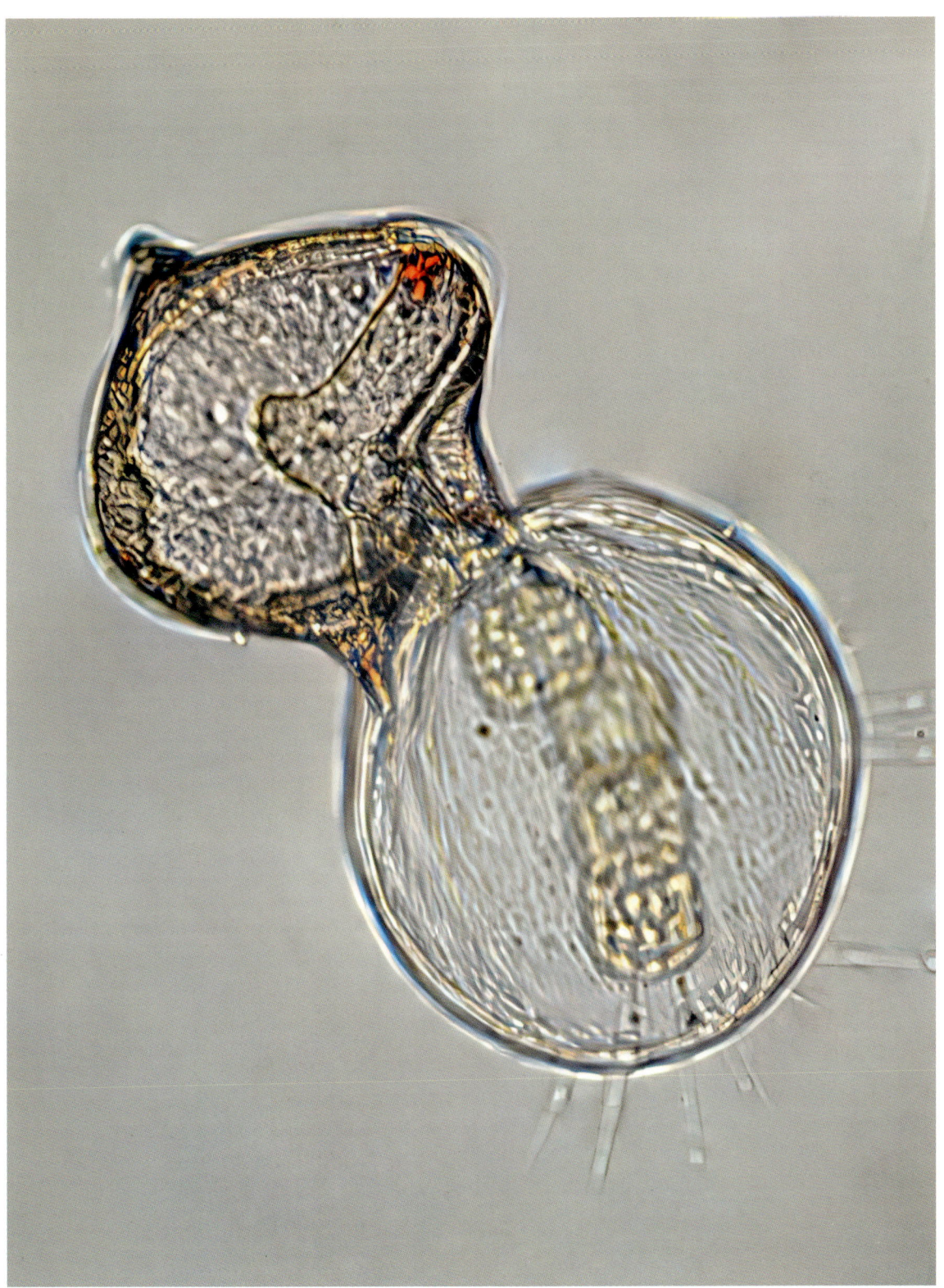

Bioluminescent dinoflagellates

Large surface blooms of dinoflagellates—hundreds to thousands of cells in every drop of seawater—emit spectacular displays of vivid, sparkling, neon-blue light that can illuminate the surface waters of entire bays or patches of the ocean during the night. Water movements, such as rolling waves and where the sea laps the shore, trigger these flashes of light. Dinoflagellates are responsible for much of the bioluminescence observed on the ocean's surface.

EVOLUTION OF BIOLUMINESCENCE

Bioluminescence—the light from within a living organism—is estimated to have evolved independently among living organisms at least 40 times. It has been observed globally since the first millennium.

Only relatively few dinoflagellate species are known to be bioluminescent. There are approximately 30 photosynthetic and even fewer nonphotosynthetic species that are known to have this capability. As with all bioluminescent systems, dinoflagellate bioluminescence is unique to these organisms from both a cellular and molecular perspective.

LIGHT EMITTERS

Bioluminescence is generated by a chemical reaction in vesicles, tiny specially made membrane-bound sacs, which are 0.5-0.9 microns in diameter. These sacs are called scintillons, and were named for their ability to engage in twinkling. They are located in the peripheral cytoplasm of the cell, particularly at night, and project into the vacuole of the cell. These vesicles contain the light-emitting luciferin substrate, the luciferase enzyme, and, in some species, a luciferin-binding protein. In response to mechanical agitation, an influx of acid (protons) from the cell vacuole into the scintillons reduces its pH, which activates the luciferin-luciferase reaction to produce a flash of light. Most scintillons, around 90 percent, break down at the end of the night and are synthesized afresh the next evening.

Noctiluca scintillans, as the name aptly indicates (Latin *nox*, night and *lucere* to shine; *scintillare*, twinkle), generally forms large vivid red blooms at the ocean surface worldwide (see page 255) that are strongly bioluminescent at night. Its buoyant cells are able to migrate from deeper waters to the ocean surface. Dinoflagellate bioluminescence is thought to act as a "burglar alarm" to signal the presence and disrupt the feeding behavior of their predators.

↖ Neon-blue light emitted by a massive bioluminescent bloom of the dinoflagellate *Noctiluca scintillans* spectacularly illuminates the waters of Ralph's Bay, Tasmania, Australia.

← An atypical dinoflagellate, the very large (up to 2,000 microns in diameter) balloon-shaped cells of *Noctiluca scintillans* bloom in coastal waters worldwide.

Algal blooms

Algal blooms can be natural events or a consequence of increased nutrient loading into aquatic environments by the human race. Many natural algal blooms are crucial to the planet's health via their significant contributions to global biogeochemical cycles, while diatom blooms fuelled by upwellings of nutrient-rich water from the seafloor support commercial fisheries. However, human-mediated blooms potentially cause detrimental changes to aquatic environments and negative impacts on human health.

HARMFUL ALGAL BLOOMS

Increased primary production fuelled by nutrient enrichment is not problematic if it is consumed by herbivores and carnivores and passed along food webs. However, harmful algal blooms can occur when a proportion of the primary production is not consumed but accumulates as organic matter in the sediments of aquatic environments. Even small annual increments of organic matter over several to many years will eventually lead to progressive changes in the species composition and abundance in the algal communities.

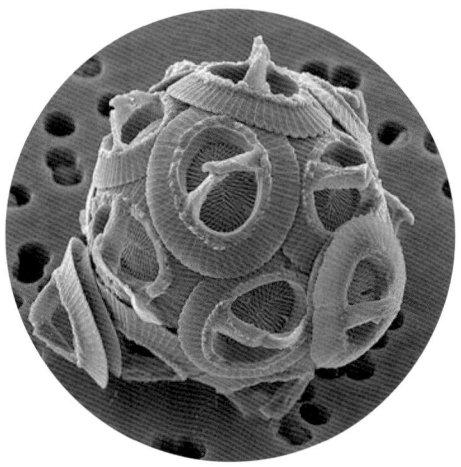

As the environmental degradation worsens, the environment selects the algal species that are best adapted to it. These tend to be "weedy" species with rapid growth rates and wide environmental tolerances, while algal species that are sensitive and intolerant to the environmental degradation are lost.

Many harmful algal blooms form in enclosed bays, estuaries, and freshwater lakes with long water residence times, as this allows the organic matter to be retained in the sediments. As the organic matter decomposes, oxygen is consumed and the water becomes hypoxic. Having evolved in hypoxic environments, cyanobacteria often form blooms in degraded freshwater environments with low dissolved oxygen levels.

Not all species cause recurrent algal blooms, just a notorious few, compared to the large number of algal species. The marked differences in the ecology and physiology between phytoplankton and macroalgal blooms require different strategies for bloom management. The microscopic cells that cause phytoplankton blooms are often so numerous that they color the water, attaining cell densities of almost 4.55 million cells per gallon (1 million cells per liter).

In the marine environment, milky blooms are caused by haptophytes, while red blooms—"red tides"—are generally caused by dinoflagellates and cryptophytes. Red water blooms in freshwater are

caused by the filamentous cyanobacterium *Planktothrix rubescens*, the euglenoid *Euglena sanguinea*, and less frequently by dinoflagellates. Conversely, macroalgal blooms are composed of attached or floating thalli, which decompose when the bloom collapses and often become stranded as stinking rotting masses on beaches. Species of cyanobacteria (*Lyngbya*), green algae (*Ulva*, see page 151; and *Cladophora*), and brown algae (*Ectocarpus*, *Hincksia*) are all common bloom-forming macroalgae.

↑   Massive nuisance blooms of sea lettuce (*Ulva*) have stranded on beaches in Brittany, France, over recent decades.

←   The highly reflective calcareous scales that cover the cells of *Gephyrocapsa oceanica* (seen in this electron microscope image) give blooms of this haptophyte species a milky appearance.

↑ Long stipes from a submarine forest of bull kelp (*Nereocystis luetkeana*) float on the sea surface during low tide in the Salish Sea, Washington state.

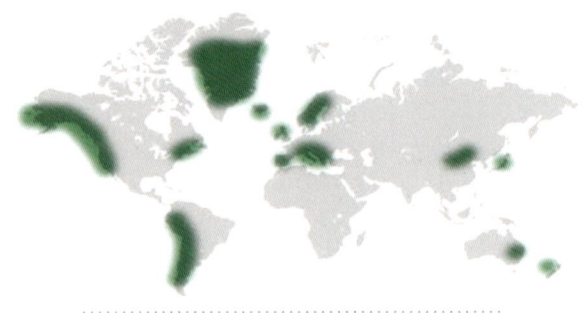

KINGDOM : Plantae
PHYLUM : Chlorophyta
CLASS : Chlorophyceae
ORDER : Chlamydomonadales
GENUS : *Chlamydomonas*
HABITAT : High alpine and polar snowfields

PHYLUM CHLOROPHYTA

Chlamydomonas nivalis

Snow alga

Unicellular snow algae live in and on the snow in alpine and polar environments, where they survive the freezing temperatures of winter, intense solar radiation, low nutrient levels, and the frequent summertime freeze-thaw cycles that may cause desiccation stress.

In summer, many of these microscopic algae—too small to be seen with the unaided eye—form striking red, pink, orange, green, golden-brown, or gray blooms on the snow surface, creating a phenomenon that has fascinated humans throughout the ages, and that, in its most common form, is still often described as "red snow." Snow algae occupy a specialized and unique habitat. They thrive in the liquid water film surrounding the ice crystals and the melting snow and only reproduce in this habitat.

At least 140 genera and 300 microalgal species are known to inhabit the snowfields of most continents. Snow is most commonly colored by the unicellular green algae—particularly *Chlamydomonas nivalis* and several other species of *Chlamydomonas*—and certain species of the closely related genera *Chloromonas* and *Chlainomonas*. These three genera are widely distributed in nature, but have relatively few snow species—only 10 of the approximately 140 species recorded for *Chloromonas* are snow dwellers. Snow algae are also known in other algal phyla: blooms of the charophyte *Mesotaenium berggrenii* color glaciers a subtle gray; the golden-brown

chrysophyte *Kremastochrysopsis* turns melting summer snow yellow; and the dinoflagellate *Borghiella pascheri* causes red snow and ice.

THE SUNSCREEN

The "blood snow" algae are unicellular green algae that produce red-resistant spores to survive the extreme environmental conditions during winter. Their green-pigmented vegetative cells synthesize and accumulate a red pigment that, in sufficient quantities, masks the green chlorophyll in the chloroplast and gives the cell a red appearance.

Chlamydomonas nivalis synthesizes unusually large amounts of the red pigment astaxanthin, which accumulates in lipid globules in the cytoplasm separate from the photosynthetic pigments found in the chloroplast. The astaxanthin acts as the alga's "sunscreen," shading and protecting the photosynthetic light-harvesting pathway in the chloroplast from the detrimental effects of the intense sunlight of the snowfields. Decreasing the amount of light striking the chloroplast potentially reduces the risk of photoinhibition (the inhibition of photosynthesis) and photodamage to the chloroplast.

→ An algal bloom of the microscopic cells of the green alga *Chlamydomonas nivalis* colors the snow red in the Dolomite Mountains, Italy.

KINGDOM	:	Chromista
PHYLUM	:	Ochrophyta
CLASS	:	Phaeophyceae
ORDER	:	Fucales
GENUS	:	*Sargassum*
HABITAT	:	Pelagic

PHYLUM OCHROPHYTA

Sargassum and the Sargasso Sea

Drifting Sargassum

Macroalgae that grow attached to the substrata are often detached by storms. These detached plants will survive if they float just under the sunlit surface of the sea, as is the case with the drifting thalli of the brown seaweeds *Sargassum fluitans* and *Sargassum natans*.

Drifting seaweeds are found in the Sargasso Sea, which is a becalmed area located in the middle of the deep ocean in the North Atlantic subtropical gyre, often said to be the only sea without a coastline. Trapped in the gyre by encircling oceanic currents, *Sargassum fluitans* and *Sargassum natans* both have pneumatocysts that float the thalli in the sunlit surface waters, forming massive floating communities that are visible in satellite imagery. They are the world's only long-term pelagic seaweeds.

This community of drifters is thought to have formed over 40 million years ago, seeded by populations in the Gulf of Mexico. The Gulf Stream now transports an estimated 1.1 million tons (1 million metric tonnes) of *Sargassum* biomass to the Sargasso Sea each spring. The *Sargassum* thalli are ecosystem engineers; these dense floating meadows effectively modifying the oceanic environment to produce rare and productive habitats in the deep ocean far from land. This valuable and unique ecosystem provides food, habitat, and nursery grounds for diverse assemblages of animals. Ten animal species are endemic, many having developed adaptations for life among the seaweed. The most iconic is the Sargasso angler fish, which has developed camouflage and modified fins to equip it for life crawling among the seaweed. The seaweed provides habitat for 145 species of invertebrates and 127 species of fish. Turtles, swordfish, sharks, and tuna are temporary residents passing through the Sargasso Sea during their migrations.

Intriguingly, life at sea has resulted in the *Sargassum* losing their ability to reproduce sexually. Instead of forming the sexual reproductive organs found in other species of *Sargassum*, the sterile thalli reproduce solely through vegetative fragmentation.

THE SARGASSUM RAFT

In 2011, a large *Sargassum* raft split from the main Sargasso raft, forming the Great Atlantic Sargassum Belt. This belt consisted of more than 22 million tons (20 million tonnes) of biomass and stretched for over 5,280 miles (8,500 km), from Mexico to the west coast of Africa. Massive and unprecedented "golden tides" have also led to periodic strandings of *Sargassum* on beaches in Brazil. In 2015, around 330 tons (300 tonnes) of *Sargassum* washed up on a Brazilian beach.

→ Thick, floating *Sargassum* mats create the unique, diverse oceanic ecosystem of the Sargasso Sea. Remote from land in the center of the North Atlantic Ocean, this ecosystem is impacted by plastic pollution.

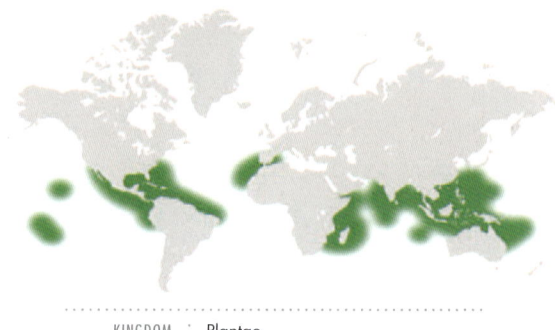

KINGDOM	:	Plantae
PHYLUM	:	Chlorophyta
CLASS	:	Ulvophyceae
ORDER	:	Bryopsidales
GENUS	:	Halimeda
HABITAT	:	Coralgal reefs to depths of around 500 ft (approximately 150 m)

PHYLUM CHLOROPHYTA

Halimeda communities

Green segmented calcareous alga

The siphonous *Halimeda* is one of the most species-rich, prevalent, and ecologically important green algal genera in tropical marine ecosystems.

Species of *Halimeda* are particularly abundant on coralgal reefs, where they are important primary producers and structural components of tropical reef systems, providing food and habitat to diverse assemblages of invertebrates and fish. Within the community, *Halimeda* commonly grows with many other siphonous green algae, most notably *Udotea* and *Penicillus*.

Halimeda is a major occupier of space on coralgal reefs, attaining a high biomass in a range of reef habitats. Various species are capable of colonizing either sand or rock substrata—*Halimeda incrassata* anchors itself into the sand with a large subterranean bulbous holdfast, while *Halimeda heteromorpha* has a felt-like holdfast that attaches it to rock. Their thalli form thick, sprawling mats, creating one of the major phototrophic communities in the reef lagoon.

Other species only thrive in deeper water. *Halimeda cryptica* grows on vertical rocky walls along deep reef slopes at 100–165 ft (30–50 m), and *Halimeda kanaloana* inhabits the sand and rubble bottoms at depths of 50–280 ft (15–85 m). In the deeper rocky escarpments of the reef, *Halimeda* grows as large draperies or as vine-like pendulous thalli.

ECOLOGICAL SUCCESS

Species of *Halimeda* have rapid growth rates, various modes of asexual and sexual reproduction, and the ability to rapidly colonize any available space, adaptations that promote the formation of lush meadows. Thalli grow from the tips, with one species (*Halimeda opuntia*) rapidly producing 28 new segments in 41 days. *Halimeda* species reproduce vegetatively, with new thalli produced by elongate subterranean rhizoids and from thallus fragments. Disturbance caused by tropical storms and hurricanes simultaneously clear space by ripping living organisms off the seafloor and producing many thallus fragments of *Halimeda* that rapidly colonize the bare space. Even white thallus fragments buried in sand can grow into green plants when exposed to light. Sexual reproduction generates zygotes that colonize bare space further afield. *Halimeda incrassata* uses these strategies to compete for space with seagrasses in Caribbean reef lagoons.

→   Luxuriant subtidal meadows of *Halimeda copiosa* carpet the rocky seafloor and provide habitat for invertebrates and fish in the Maldives, islands south of India in the tropical Indian Ocean.

KINGDOM	:	Chromista
PHYLUM	:	Ochrophyta
CLASS	:	Phaeophyceae
ORDER	:	Laminariales
GENUS	:	*Postelsia*
HABITAT	:	Mid-intertidal zone on very exposed rocky shores

PHYLUM OCHROPHYTA

Postelsia palmaeformis

Sea palm

The sea palm is restricted in its ecological distribution to the mid shore on rocky coasts that are exposed to extreme wave action. Thalli of this diminutive kelp are only 20–30 in (50–75 cm) tall.

Its morphology—a crown of leaflike blades atop a stemlike stipe that is attached to the substratum by a branched holdfast—resembles, although much reduced in size, its close relative, the giant bull kelp. Unlike most macroalgae, the sea palm stands upright like a tree when it is left high and dry at low tide. This surf zone specialist maintains this posture throughout the tidal cycle except when hit by an incoming wave. With each wave, the flexible, mostly hollow stipe bends until it is almost prostrate, only to spring back to its upright position once the wave has passed.

Although exceptionally robust in the face of crashing waves, the sea palm has limited distribution on the seashore. Its upper limit is determined largely by the temperature and desiccation experienced during exposure at low tide, while its lower distribution limit is set by its inability to survive the lower light levels found during longer periods of submersion on the lower shore. Consequently, the sea palm is adapted to the mid shore, where there are higher light levels without the risk of desiccation or extreme temperatures.

Often outcompeted on the mid shore for space by its primary competitor, the California mussel, the sea palm depends on winter storms and water-borne logs ripping out patches of mussels to clear space for the young kelps to colonize in following early spring. Within the cleared spaces, the sea palm typically grows into dense, crowded stands reminiscent of dwarf palm forests, although some isolated individuals also grow on the shore. This is a finely balanced system. The sea palm appears to need the mussels to exclude the turf algae that also grow on the mid shore and in whose presence the kelp's microscopic spores and young thalli cannot attach to the rocks. This probably explains why the sea palm only grows on coasts exposed to extreme wave action. It needs the disturbance caused by the extreme wave action each winter to clear space on the rocks on which it can grow.

→ Resembling a miniature palm tree, the sea palm (*Postelsia palmaeformis*) competes for space with mussels and turf algae on wave-exposed rocky shores, from cool temperate California to British Columbia on the North American west coast.

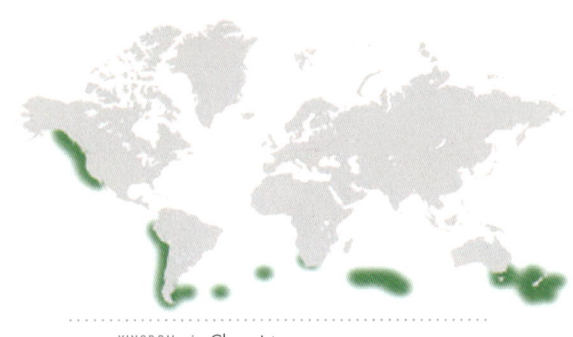

KINGDOM	Chromista
PHYLUM	Ochrophyta
CLASS	Phaeophyceae
ORDER	Laminariales
GENUS	*Macrocystis*
HABITAT	Subtidal zone on rocky shores

PHYLUM OCHROPHYTA

Macrocystis pyrifera

Giant kelp

There are three giant kelps that form the canopy species in the spectacular giant kelp forests of the cool temperate northeast Pacific coasts: the aptly named giant kelp, the bull kelp, and the elk kelp.

Each of these species grows to more than 100 ft (approximately 30 m) in length, with geographical ranges that overlap on the cool temperate northeast Pacific coast. The giant kelp (*Macrocystis pyrifera*) and the bull kelp (*Nereocystis luetkeana*) grow in mixed submarine forests to maximum depths of 100–115 ft (30–35 m). *Macrocystis pyrifera* is a perennial species whereas the bull kelp is an annual species that is only present from March to October. Elk kelp (*Pelagophycus porra*) forests grow seaward of the *Macrocystis* forests, at depths of 65–165 ft (20–50 m).

Growing from the dimly lit seafloor, the kelp thalli are buoyed upright by gas-filled pneumatocysts, with their fronds spreading across the sunlit surface waters. The pneumatocysts give the kelps the strategic advantage of positioning the fronds where they will typically absorb 80 percent of the light falling on the uppermost 3 ft (approximately 1 m) of the sea.

The three giant kelp species are the ecosystem engineers in these communities. They alter the light regime and increase the shelter within the community and, in doing so,

determine the structure of the understory vegetation. The kelp forests also slow longshore currents by around 30 percent, and this relatively stagnant water, along with the kelp fronds, provides many fish and invertebrates with a safe refuge from predators.

THE GIANT KELP COMMUNITY

Californian *Macrocystis* forests typically have a rich and diverse understory, which often grows at different levels under the kelp canopy. In the uppermost layer, the thalli of the smaller kelps—*Pterygophora californica* and *Eisenia arborea*—stand erect, supported by their stiff stipes. Below this is a layer of prostrate large brown algae lying on the seafloor that includes the kelp *Laminaria farlowii* and the bladder chain kelp, *Stephanocystis osmundacea* (the latter species, with its misleading common name, is a fucoid, not a true kelp). The prostrate kelps are surrounded on the substratum by areas of dense turf that are dominated by articulated coralline red algae and bushy red algae. Encrusting coralline red algae grow in areas where sea urchins have eaten the turfs.

→   Light attenuates with depth in the three-dimensional giant kelp (*Macrocystis pyrifera*) forest that provides habitat for other algae, fish, and invertebrates at Santa Barbara Island, California, on the cool temperate North American west coast.

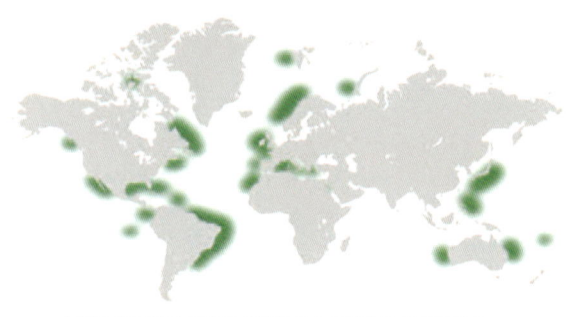

KINGDOM	Plantae
PHYLUM	Rhodophyta
CLASS	Florideophyceae
ORDER	Corallinales
GENUS	*Lithothamnion*
HABITAT	Soft bottoms in upper to deep subtidal zone

LITHOTHAMNION CORALLIOIDES

Rhodolith communities

Deep sea algae

Rhodolith—or maerl—beds are communities of free-living nodules of encrusting coralline red algae that have accumulated over thousands of years on the seabed, creating fragile, structured ecosystems.

These ecosystems are bioconcretions, a structure formed of solid particles created by living organisms. The physical structure of the rhodolith beds is formed from the calcareous nodules or bodies of the encrusting coralline red algae. Rhodolith communities form extensive beds over broad latitudinal and depth ranges worldwide, occurring from the lower intertidal zone to water depths of around 500 ft (150 m). It appears that light and salinity are key factors in determining the depth and biogeographic distribution of the rhodoliths, but water motion in the form of waves and currents is also important, as it creates and keeps the rounded or branched thalli rolling on the seafloor in an unattached and unburied state.

Rhodolith communities are particularly well developed in the Mediterranean Sea, where they are estimated to cover more than 1,000 sq miles (2,700 sq km) of the seabed to a height of 16 ft (5 m).

Encrusting coralline red algae are ecologically important as ecosystem engineers in these communities. They build a high-diversity habitat in the form of a three-dimensional lattice framework, with the spaces in the lattice providing microhabitats for diverse assemblages of invertebrate animals, including sponges, jellyfish, lace corals (bryozoans), sea squirts (tunicates),

marine bristle worms (polychaetes), and mollusks. Other algae and invertebrates grow on the surface of the rhodoliths; unusual and endemic algal species have been reported in many rhodolith beds, from tropical to polar regions.

FOUNDATION SPECIES

The foundation species in rhodolith beds vary geographically. European, Mediterranean, and Gulf of Californian rhodolith beds are dominated by different rhodolith species.

In European rhodolith communities, some 70 to 90 noncalcified macroalgal species have been recorded, with several macroalgal (fleshy) species more or less confined to these rhodolith beds. Red algal species are the most abundant and dominant, which is consistent with reports of red algal dominance in rhodolith beds worldwide; the fleshy macroalgae are seasonal and have the highest abundance in Europe in summer, and in the Gulf of California in winter.

In Europe, maerl beds are threatened by the large-scale human extraction for fertilizer, aquaculture facilities, sewage discharge, and bivalve dredging. The European Union has legally protected maerl beds from further human exploitation.

→ The shelflike red seaweed *Peyssonnelia inamoena* and the brown seaweed *Dictyota* (lower left) are common in rhodolith communities growing at depths of 200 ft (66 m) in the Gulf of Mexico.

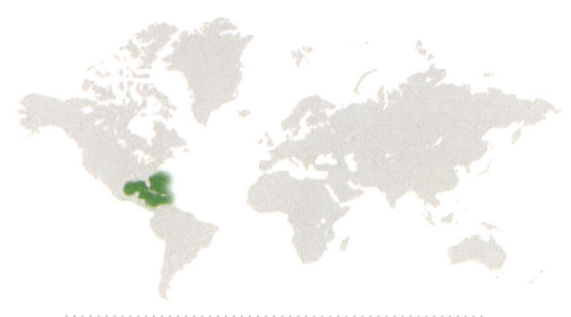

Symbiodinium microadriaticum

Symbionts

KINGDOM	Chromista
PHYLUM	Miozoa
CLASS	Dinophyceae
ORDER	Suessiales
GENUS	*Symbiodinium*
HABITAT	Planktonic or endosymbiont in coral polyps

The 2014 to 2017 global mass coral bleaching event—the third in the history of the planet—grabbed headlines around the world. The first in 1998 and the second in 2010 were of relatively brief duration.

Coral bleaching events are caused by severe, rapid, or prolonged thermal stress in response to elevated sea surface temperatures, and can ultimately lead to massive coral mortality and reef degradation. The coral bleaches white when either the symbiont's photosynthetic pigments are lost or the symbiont is expelled from the coral polyp. The coral polyp relies on its photosynthetic symbiont—a dinoflagellate, commonly known as zooxanthellae—for its nutrition and will die if it is not recolonized by the symbionts.

In 1962, *Symbiodinium microadriaticum* was described living in the upside-down jellyfish *Cassiopea xamachana*. This jellyfish lies on the shallow seafloor with its saucer-shaped bell pointing downward and its eight oral arms directed up toward the sun. This habit maximizes the photosynthetic rate of the symbionts that reside in the oral arms. For the next two decades, it was assumed that only one species of zooxanthellae—*Symbiodinium microadriaticum*—inhabited corals and other marine invertebrates. The symbionts of these marine animals appeared identical and had been little studied by science.

The revelation that *Symbiodinium* comprised many species began in 1987 with electron microscopy studies and has continued until now, with DNA sequencing studies recognizing at least 17 new *Symbiodinium* species—some the symbionts of coral species—and identifying nine evolutionary divergent lineages that are thought to comprise a multitude of as yet undescribed species.

The typical dinoflagellate motile cells of *Symbiodinium microadriaticum* are small and covered by delicate thecal plates, with the upper half slightly larger than the lower half of the cell. *Symbiodinium* reside in a complex host and symbiont-derived, membrane-bound cytoplasmic organelle known as the symbiosome. Zooxanthellae also inhabit sponges, sea anemones, giant clams, and nudibranchs, as well as unicellular ciliates and forams. Coral–*Symbiodinium* associations are not reciprocally exclusive but flexible. Zooxanthellae are frequently released from healthy corals and new zooxanthellae taken into the coral polyp from a free-living symbiont pool in the environment.

→ Severe thermal stress bleaches corals such as this staghorn coral (center), revealing the coral's white skeleton following the loss of its symbionts. Under moderate thermal stress, however, corals either partially bleach, like the branching corals (background), or produce fluorescent pigments like the lavender coral (foreground) to shade the symbionts.

Dinophysis caudata

Predator

KINGDOM	:	Chromista
PHYLUM	:	Miozoa
CLASS	:	Dinophyceae
ORDER	:	Dinophysiales
GENUS	:	*Dinophysis*
HABITAT	:	A coastal marine plankton

In 2006 (and after many unsuccessful attempts), a species of the dinoflagellate genus *Dinophysis* was grown in laboratory cultures for the first time. This was achieved by including its ciliate prey in the cultures.

The discovery uncovered a remarkable scenario: the *Dinophysis* had harvested its plastids from its ciliate prey, which in turn had harvested the same plastids from its algal prey. It is quite bizarre that, having consumed the cell of the cryptophyte alga (*Teleaulax amphioxeia*), the ciliate (*Myrionecta rubra*) digested the cell but retained the cryptophyte's fully functional plastids in its cell, and that the process was then repeated when the *Dinophysis* cell preyed on the ciliate. The cryptophyte and ciliate are both abundant, and can form red-colored blooms in coastal marine ecosystems. Although both photosynthetic and heterotrophic species are common in the dinoflagellates, it had been generally assumed that dinoflagellate species with plastids were phototrophs. The discovery that *Dinophysis* needed to harvest plastids from their prey for survival was a surprise.

Interestingly, the phototrophic *Dinophysis* species are not known to possess typical dinoflagellate plastids (with the pigment peridinin and three enveloping membranes). Instead, they only have cryptophyte plastids (with the pigment phycoerythrin and two enveloping membranes). *Teleaulax amphioxeia* is a common food of several unicellular organisms.

When consumed, its plastids remain photosynthetically active, sustaining the growth of the ciliate and *Dinophysis caudata*. If deprived of this important food source, *Dinophysis caudata* only remains photosynthetically active for two months, after which it gradually dies. If, however, it is fed the ciliate after two months (and before it dies), *Dinophysis caudata* acquires new plastids and resumes photosynthetic activity.

During feeding, *Dinophysis caudata* actively swims around and makes physical contact with its swimming ciliate prey. The dinoflagellate extends a feeding tube (peduncle) from its cell that it inserts into the ciliate, immobilizing it instantaneously and causing the shedding of the prey's cilia (suggesting that a toxin is involved). The predator then actively ingests the ciliate's cell contents and its stolen red plastids are soon visible inside the dinoflagellate cell. Curiously, *Dinophysis caudata* does not prey directly on the cryptophyte—it only harvests the cryptophyte plastids from the ciliate!

→   This light microscope image shows a feeding sequence by the large cell of the predatory dinoflagellate *Dinophysis caudata* (with sails) that is sucking through its feeding tube and transferring into its cell the protoplast, including the red plastids, of its smaller spherical ciliate prey (center cell pair). This leaves the ciliate's cell almost colorless (top right cell pair).

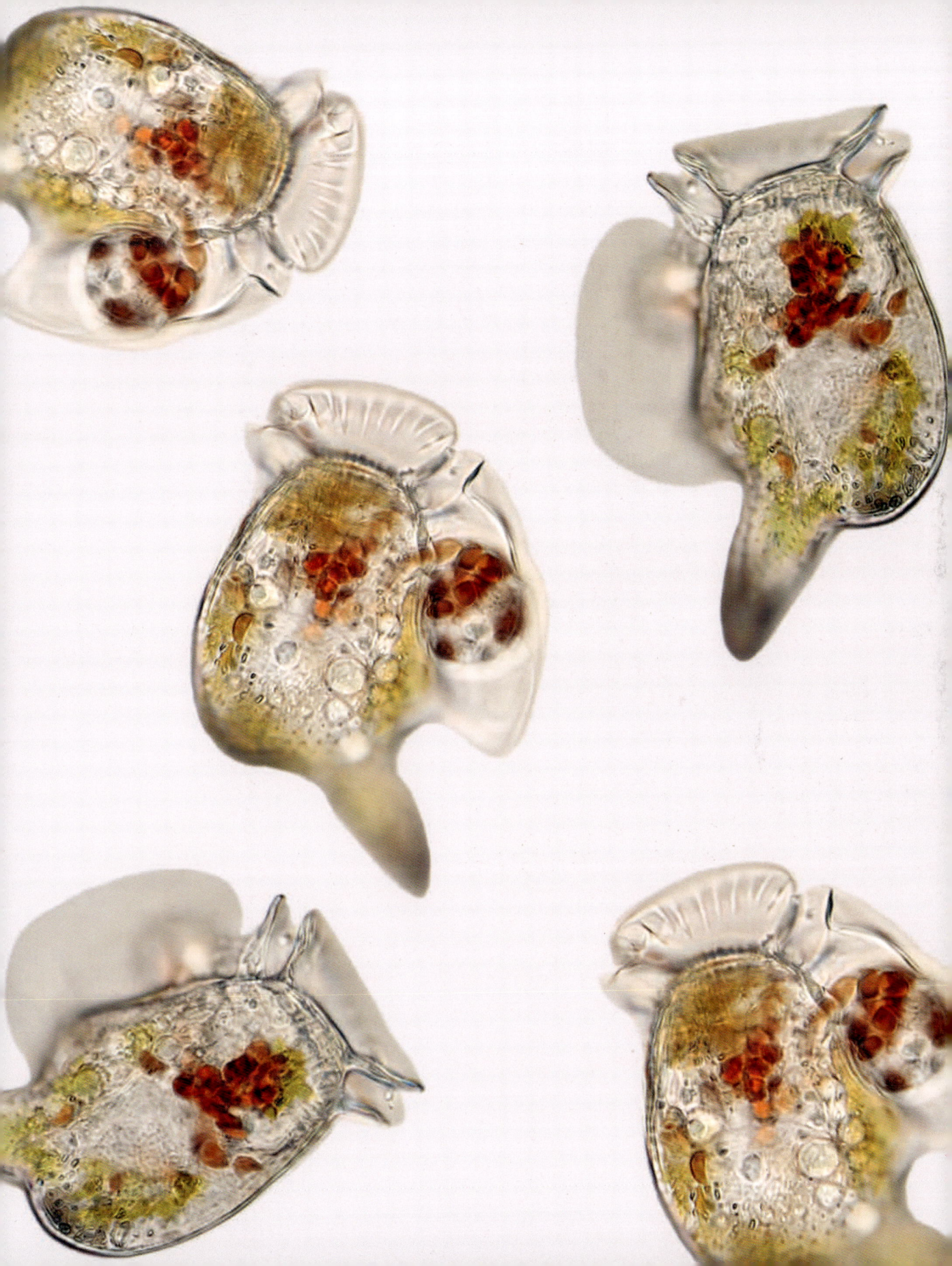

PHYLUM CYANOBACTERIA

Trichodesmium spp.

Sea sawdust

KINGDOM	:	Eubacteria
PHYLUM	:	Cyanobacteria
CLASS	:	Cyanophyceae
ORDER	:	Oscillatoriales
GENUS	:	*Trichodesmium*
HABITAT	:	Oceanic plankton

Tropical oceans are blue—the color of marine nutrient deserts. Despite its low-nutrient oceanic habitat, species of the planktonic *Trichodesmium* form massive recurrent blooms that are visible from outer space and have intrigued naturalists, biologists, and mariners for more than a century.

Trichodesmium is found in tropical and subtropical oceans worldwide, where its microscopic single filaments or bundles of filaments live suspended in the sunlit surface waters. Bundles of filaments also form enormous accumulations that float on the sea surface over many hundreds of square miles. These surface accumulations resemble sawdust and are commonly known as sea sawdust.

Trichodesmium blooms are driven by the ocean currents and winds onto coastlines, where the sea sawdust often fouls the beaches with a reddish-brown slick. The Red Sea is reported to take its name from sea sawdust blooms.

Paradoxically, the photosynthetic *Trichodesmium* achieves an extraordinarily high biomass in the nutrient-poor oceans. This indicates that *Trichodesmium* is uniquely adapted to an environment in which many other phytoplankton species struggle to survive. The key to this ecological success is its ability to bypass the nutrient limitation to phytoplankton growth in tropical seas. *Trichodesmium* has the capacity to use atmospheric nitrogen gas in the chemical process of nitrogen fixation.

In this process, nitrogen gas is converted into ammonia, which is used to synthesize protein from carbohydrate. It also employs gas vesicles, cell growth, and carbohydrate ballasting for buoyancy regulation and vertical migrations. By sinking and rising daily through the water column, the filaments of *Trichodesmium* can strip the low levels of phosphorus, iron, and other nutrients from a greater volume of seawater, thereby optimizing its growth. Buoyancy also ensures that *Trichodesmium* remains in the sunlit surface layers of the sea, optimizing the light levels for photosynthesis.

→   Visible to the unaided eye, millions of microscopic filaments of the pelagic cyanobacterium *Trichodesmium* aggregate to form reddish-brown sea sawdust that floats on the surface of tropical oceans, including at Magnetic Island, Great Barrier Reef, Australia.

ALGAE &
HUMANS

Globally relevant algae

The algae have provided us with food since prehistoric times, and today they are recognized worldwide as valuable sources of chemical compounds that have applications in human and animal foods. They are also used in the energy sector in the production of biodiesel and bioethanol, and in pharmaceuticals, cosmeceuticals, and nutraceuticals. Less well known are the important roles algae play in the biogeochemical cycles that sustain life on Earth.

GLOBAL BIOGEOCHEMICAL CYCLES

As phototrophic organisms, the algae drive the chemical interactions that link the hydrosphere (aquatic environments), atmosphere, lithosphere (rocks, soil), and biosphere (living organisms). Algal species that maintain a large population in the vast contemporary oceans—the oceans cover 70 percent of the Earth' surface—are particularly important in the global carbon, sulfur, nitrogen, and silicon cycles. Key algal drivers in this regard are the haptophytes (*Emiliania huxleyi*, *Gephyrocapsa oceanica*, and three species of *Phaeocystis*); the diatoms (including the polar diatom *Fragilariopsis*); the cyanobacterium *Trichodesmium*; the siphonous green alga *Halimeda*; and encrusting coralline red algae.

    Emiliania huxleyi (see page 260) is the most abundant haptophyte species in today's Atlantic, Pacific, and Indian Oceans. In the North Atlantic Ocean, its summer to early fall blooms cover more than 95,000 sq miles (250,000 sq km). *Gephyrocapsa oceanica* is the dominant haptophyte species in tropical and subtropical latitudes, blooming in midwinter to spring. These two species construct coccoliths—exquisite calcium carbonate scales that cover their cells. Another haptophyte genus, *Phaeocystis*, lacks coccoliths; three species in this genus have mucilaginous colonies that bloom in the spring and summer, either in the North Sea, the Arctic Ocean, or the Southern Ocean.

    Macroalgae can also achieve large population sizes, with some species of *Halimeda* forming extensive meadows that can cover many miles of the seafloor on tropical coasts and coralgal reefs. The dead calcified plants from these meadows are major contributors to the formation of carbonate sand sediments and limestone rock.

← A satellite image of large natural blooms of *Emiliania huxleyi* (blueish-green swirls) in the North Atlantic Ocean near Iceland.

A

SEATTLE

B

SAN FRANCISCO

C

LOS ANGELES

PACIFIC OCEAN

THE GLOBAL CARBON CYCLE

The majority of algal species are oxygenic phototrophs, and are therefore important components of the global carbon cycle. They generate their energy through photosynthesis, while some also use photosynthesis to drive the calcification of their thalli. Photosynthesis and calcification both transfer large quantities of carbon dioxide from the atmosphere to the biosphere. Photosynthesis fixes carbon dioxide into the organic compound glucose, and calcification fixes carbon dioxide into the calcifying compound calcium carbonate. It is estimated that the photosynthesis of marine phototrophs (mostly phytoplankton) fixes 46 percent of the annual global primary production, a global measure of the amount of carbon fixed each year minus losses from respiration.

This is a significant contribution, slightly less than half of the total for the planet; the other 54 percent of the global primary production is fixed by the land plants.

Not all areas of marine environment are equally productive. Highest marine primary production occurs where phytoplankton thrive in the cold, nutrient-rich waters of polar seas in the North Atlantic, North Pacific, and Southern Oceans, and in areas of upwellings that form both in a band around the

↑   Natural diatom blooms (dark-green swirls), seen in this satellite image of the California Current system on the North American west coast, are fueled by upwelling from the seafloor.

equator in the Atlantic, Pacific, and Indian Oceans, and along the coastlines of continents, particularly the west coasts of North and South America and Africa. Upwellings transport cold, nutrient-rich water from the seafloor to the sea surface. The lowest primary production is evident in the middle of ocean basins far from land.

Phototrophs are the only living organisms on Earth able to remove carbon dioxide from the atmosphere. It is also worth noting that a proportion of the carbon in marine primary production is transported to the deep ocean floor by sinking dead phytoplankton cells. This carbon is sequestered in the hydrosphere, keeping it locked away from the atmosphere unless it is transported to the ocean surface by an upwelling.

DIATOM PRIMARY PRODUCTION

The ecologically successful diatoms contribute 20 percent of the fixed carbon to global primary production, a large proportion of which is passed along grazing food webs and is important in sustaining fisheries production. Relative to the other phytoplankton phyla, diatoms remove more carbon dioxide from the atmosphere and sequester more carbon on the ocean floor due to their large cell sizes, high growth rates, and their ability to form blooms. They are also major producers of "new" organic carbon, which is fueled by upwellings of nutrient-rich water from the ocean depths, rather than nutrients recycled from organic matter in the surface waters.

Coastal upwelling
Warm, low-nutrient surface waters, blown away from the coastline and out to sea by offshore winds, are replaced by upwellings of cooler, nutrient-rich water from the seafloor.

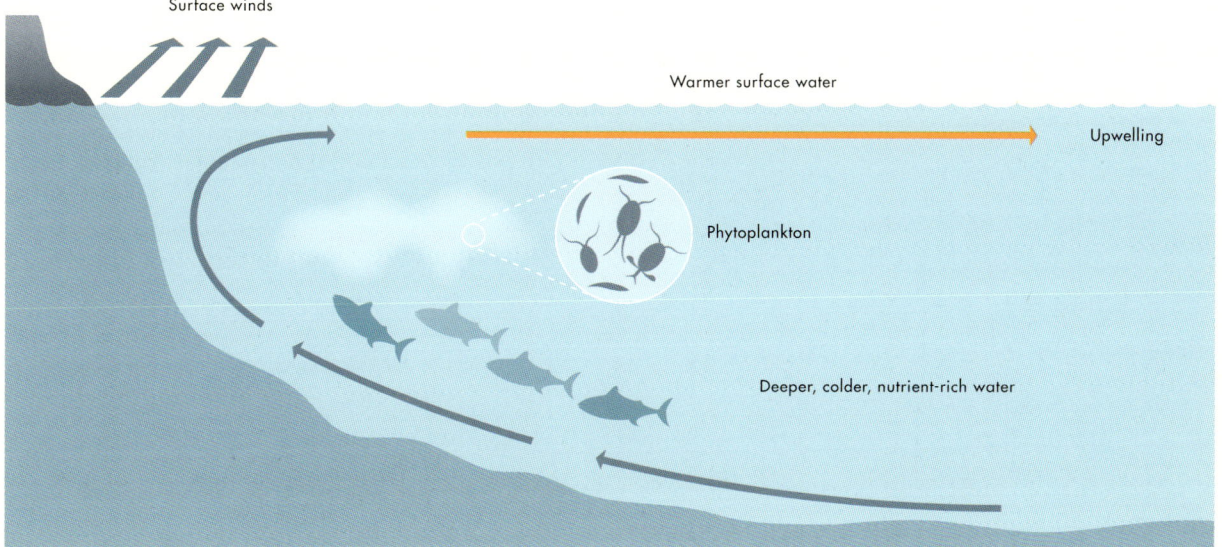

Surface winds

Warmer surface water

Upwelling

Phytoplankton

Deeper, colder, nutrient-rich water

ALGAL CALCIFICATION

By virtue of their huge biomass and their production of calcium carbonate, calcareous algae—such as the coralline red algae, and the green alga *Halimeda*—contribute to the global carbon cycle during calcification. The process of calcification relies on carbon dioxide from the atmosphere dissolving in seawater to form carbonate ions, which bind to calcium ions and precipitate as calcium carbonate onto or between the algal cell walls. The calcification of reef-building coral polyps also depends on algae, in this case the coral polyp's symbiotic dinoflagellates that supply the energy for the calcification process.

When calcareous algae die, their calcified thalli are deposited on the seafloor as carbonate sediments. Here, they can form limestone rock, which is a long-term carbon sink in the lithosphere. Haptophytes were abundant during the late Cretaceous period (63–95 million years ago), and their microscopic, scalelike coccoliths constructed extensive limestone and chalk cliffs across Europe, including the iconic White Cliffs of Dover in England. Carbon in these calcium carbonate cliffs has been locked up for more than 63 million years.

Rhodolith beds and *Halimeda* bioherms also lock up the carbon in the calcified algal thalli. Large, thick rhodolith beds are found on the seafloor worldwide, from the lower intertidal zone to a depth of almost 500 ft (150 m). From the late Miocene (22 million years ago), dead *Halimeda* thalli have accumulated in the northern Great Barrier Reef lagoon as thick, moundlike bioherms, which now reach a height

← Huge drifts of foam formed from mucilage originating from the colonies in a *Phaeocystis globosa* bloom are washed ashore onto beaches.

→ This light microscope image reveals thousands of small cells embedded in copious mucilage in a large (up to 0.8 in (2 cm) in diameter) spherical *Phaeocystis* colony.

of almost 165 ft (50 m) and cover 2,350 sq miles (6,095 sq km) of the seabed. *Halimeda* mounds are also found in Indonesia, the Timor Sea, and the Caribbean Sea.

SULFUR CYCLE

Many marine algae, particularly the bloom-forming phytoplankton species, also play a key role in the global sulfur cycle, as they emit the volatile compound dimethyl sulfide into the seawater. This chemical compound has important physiological functions in the phytoplankton cells, regulating the osmotic pressure, as an antioxidant for scavenging of reactive oxygen species, and for cryoprotection. Blooms of *Emiliania huxleyi* and *Phaeocystis* release large quantities of dimethyl sulfide, which is thought to play a major role in cooling the Earth's climate.

Once ventilated from the seawater into the atmosphere, dimethyl sulfide is photo-oxidized into sulfate aerosols, which can cool the climate directly by scattering solar radiation and indirectly by seeding cloud formation (clouds reflect a large proportion of solar radiation back into outer space).

Dimethyl sulfide contributes approximately half of the total sulfur flux released by living organisms into the atmosphere, and this is sourced mainly from the marine environment. The Antarctic Ocean may contribute 10–30 percent of the global emission of this compound, despite covering just 6 percent of the total area of the world's oceans.

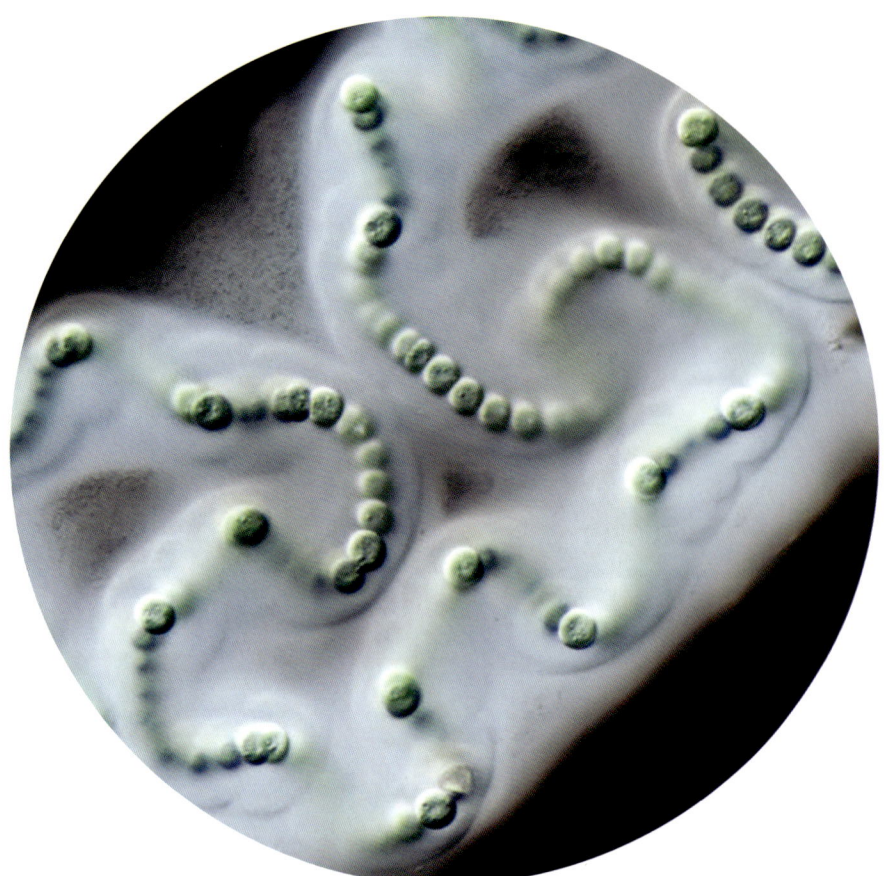

← A light microscope image of filaments of spherical cells of *Nostoc* (from the English word "nostril") embedded in copious mucilage, a striking feature that probably gave rise to the Dutch word "snot."

→ A terraced Indonesian rice paddy.

GLOBAL NITROGEN CYCLE

Among the algae, only the cyanobacteria are capable of the exclusively bacterial process of nitrogen fixation. Atmospheric nitrogen gas is converted into ammonia that the bacteria use to synthesize protein from carbohydrate. The marine cyanobacterium *Trichodesmium* (see page 227) is a significant contributor to the marine nitrogen cycle. In 1982, *Trichodesmium* was estimated to fix almost 6 million tons (5.4 million tonnes) of nitrogen a year. This figure was based on the occurrence of 20 moderately sized ocean blooms covering just over 7,700 sq miles (20,000 sq km), each of which persisted for 15 days. However, with the data from numerous subsequent studies, it is now estimated that *Trichodesmium* fixes 66–88 million tons

(60–80 million tonnes) of nitrogen each year, which accounts for almost half of the nitrogen fixed in the marine environment. The cyanobacterial symbionts living in the cells of the oceanic diatoms also contribute significant quantities of fixed nitrogen to the global nitrogen cycle. This nitrogen fixed by nitrogen fixation is "new" to the marine environment, where it enriches the fertility of the nutrient poor ocean.

The filamentous cyanobacterium *Nostoc* is another prevalent nitrogen-fixer. *Nostoc* is a common inhabitant in rice paddies—the most widespread tropical and subtropical wetlands on Earth—where it increases soil fertility and the crop yields of the rice plants.

Food and food additives

Since antiquity, many coastal peoples around the world have foraged for seaweeds and eaten them as nutritious foods. Excavations of hearths in the dwellings occupied by ancient humans in Chile more than 13,000 years ago revealed the remains of eight species of seaweeds mixed with other plant foods. The ancient people of Japan discarded pieces of seaweed, shells, and fish bones in middens between 2,300 and 8,000 years ago.

TRADITIONAL DIETS

In the present day, seaweeds are most widely eaten in Asian countries. Seaweeds, rice, and fish are the mainstays of a traditional Japanese diet. Some Japanese seaweeds are becoming familiar in the West: the red alga commonly known as nori (*Neopryopia yezoensis,* see page 262) is the delicious wrap around sushi rolls; konbu (pronounced "kombu"; *Saccharina japonica,* see page 264) is used in savory dashi stock; and wakame (*Undaria pinnatifida*) is added to salads and soups.

Seaweeds also have a history as a food source in many countries in Europe. In southwest Ireland, seaweed fragments have been found among the shoreline gatherings of Mesolithic humans who lived around 4,500 years ago. In present-day Ireland, the Irish moss (the red seaweed *Chondrus crispus,* see page 226) is used instead of gelatine to turn heated flavored milk into a pudding; while in Scotland and Wales, the laver (the red seaweed *Porphyra,* see page 158) is mixed with oatmeal, made into patties, and fried.

Another red seaweed, dulse (*Palmaria palmata*), is widely eaten in Europe, as well as in the states of New England and Maine, and the Canadian Maritime Provinces on the eastern seaboard of North America. Dulse has a piquant, salty flavor when added to sauces, mayonnaise, and soups, and is often cooked with salted cod or other fish, potatoes, and butter. In Ireland and eastern North America, dried dulse is eaten as a chewy snack.

NUTRIENT PROFILES

Seaweeds are a nutritious food source, containing proteins, dietary fibers, carbohydrates, unsaturated and saturated fatty acids, minerals (calcium, potassium, iron, zinc, and iodine), and vitamins A, B_1, B_2, B_9, B_{12}, C, D, E, and K. The nutrient levels in seaweeds vary markedly among the species and with the season, but some contain from 10 to 100 times more minerals and

vitamins per unit dry weight than foods derived from the land plants or animals. Generally, protein levels are highest in the red seaweeds, such as nori and laver (a maximum of 50 percent dry weight), compared to moderate amounts in the green seaweeds (a maximum of 32.7 percent dry weight), and low amounts in the brown seaweeds (3–15 percent dry weight).

Seaweeds are also rich in natural antioxidants, which can prevent cellular deterioration from oxidative stress conditions and help strengthen the immune system, therefore combating inflammatory and age-related diseases. In addition, chemical compounds in seaweeds such as pigments, polyphenols, sulfated complex carbohydrates, lipids, proteins, and organic acids potentially have antibacterial, antiviral, antitumor, and antifungal properties.

↑   Sushi rolls have various fillings surrounded by vinegared rice and a dried black nori (seaweed) wrap.

→   Dulse is wild-harvested from rocks on cool temperate North Atlantic coasts.

←   The kelp wakame is harvested for food in Argentina.

SEAWEED AQUACULTURE

Over the past few decades a large international seaweed industry has grown, much of which is centered on farming seaweed species in the sea. Aquaculture farms now produce more than 33 million tons (30 million tonnes) of seaweeds a year, worth US$11.7 billion, compared to wild harvests of just over 580,000 tons (525,000 tonnes). China, Indonesia, the Philippines, Korea, and Japan produce 97 percent of aquacultured seaweeds. Tanzania, Kenya, Malaysia, Chile, the states of New England and Maine in the United States, and the

Canadian Maritime Provinces have small aquaculture industries. In Nova Scotia, Canada, the Irish moss has been farmed for decades as an edible crop, in extensive land-based tank farms.

More than 35 species of seaweeds are currently cultivated commercially. The most important cultivated seaweeds in terms of biomass are the red algal genera *Eucheuma*, *Kappaphycus* (see page 272), and *Gracilaria* (and its close relatives), which are farmed for their complex cell wall carbohydrates, and the kelps konbu and wakame, the red seaweed nori; and large brown

(fucoid) alga, hijiki (*Sargassum fusiforme*), which are farmed for human food. In Japan, the demand for the popular hijiki greatly outstrips domestic supply. Each year, Japan harvests 8,800–11,000 tons (8,000–10,000 tonnes) of hijiki and imports an additional 50,000 tons (46,000 tonnes) from China and Korea.

↑   Visible at low tide, the seaweed farms on Zanzibar cultivate plants of the red, green, and brown strains of *Kappaphycus* on ropes strung between stakes.

↖   Women on the tropical island of Zanzibar, East Africa, tie small pieces of the red seaweed *Kappaphycus* onto ropes which are then "planted" in the sea.

NORI AQUACULTURE

The Japanese have farmed the red seaweed nori since the 1650s. It grew in estuaries during the fall and winter but disappeared from this habitat each summer. Each September for three centuries, the nori fishermen pushed bamboo twigs or, since the 1920s, poles on which nets were strung, into the muddy estuaries and waited for nature to seed the next nori crop. Once in every five years or so, thick bands of tiny purple specks mysteriously appeared in early November on the twigs or nets and, over the next seven weeks, grew into a crop of large, leaflike nori plants. However, in most years, the nori crop was average or it failed completely, which kept the nori fishermen poor. No one knew why in some years few or no tiny plants grew on the twigs or nets.

In 1949 and half a world away in England, Dr. Kathleen Drew solved the nori mystery that had perplexed other phycologists for the previous 70 years,

and the nori fishermen for three centuries. In her laboratory, she discovered that the spores of the leaflike laver, a close relative to nori, grew as pink spots in the limey matrix of oyster shells. Dr. Drew communicated these results to her Japanese colleague, Professor Sokichi Segawa, who informed his close friend, the fisheries engineer, Mr. Fusao Ōta. It took Mr. Ōta and the nori fishermen three years to commercialize her discovery. Over the summer, they cultured the spores from the leaflike nori with oyster shells in large rectangular concrete seawater ponds on land. In the early fall, when the pink spots on the oyster shells released a second type of spores, the nori fishermen slowly rolled their nets through the ponds. The spores settled on the nets, which were then strung between poles in the estuary. Over the next two months, large, leaflike nori grew on the nets. Once the technique for seeding of the nets had been developed, nori crop yields soared and crop failures declined.

Her gift, which lifted the nori fishermen out of poverty, has never been forgotten. Dr. Drew died in 1957, aged 55 years. On April 14, 1963, Mr. Ōta and the nori fishermen, with her proud husband, Emeritus Professor Henry Wright Baker in attendance, unveiled her memorial in the sacred grounds of a Shinto shrine in western Japan. On this day every year since 1963, the nori fishermen have assembled at the memorial to pay her homage.

Japan's nori industry is currently worth around US$850 million annually to the Japanese economy. People worldwide have benefitted from Dr. Drew's discovery that enabled nori to be grown on a commercial scale.

← Despite never having visited Japan, Dr. Kathleen Drew made a scientific discovery that saved Japan's nori industry.

→ A Japanese woodblock dated circa 1848/1852, depicting women gathering nori near Edo (now Tokyo). This delicacy is sold in Asakusa as the famous *Edo-meisan* (speciality of Edo) Asakusa nori.

MICROALGAL AQUACULTURE

Like the seaweeds, microalgae are an excellent source of proteins, polyunsaturated fatty acids, valuable pigments, and vitamins, and are often used in nutritional supplements for humans and animals. The composition of microalgae varies markedly, depending on the species and the growth conditions, but the green freshwater algae *Chlorella* and *Scenedesmus* and the cyanobacterium *Spirulina* are very rich sources of protein.

Microalgal cultures are used widely in the aquaculture of marine animals, particularly crustaceans, and mollusks. Oysters, scallops, and clams filter feed on microalgae throughout their lives whereas shrimp and prawns only feed on them during their early larval stages. The unicellular marine golden-brown alga *Nannochloropsis* is a highly nutritious source of omega-3 polyunsaturated fatty acids, while the green alga *Tetraselmis* is one of the many microalgal species used to feed the larvae of commercially important oysters, clams, and abalone.

Some microalgal species are aquacultured for specific chemical compounds, including the red pigment astaxanthin, which is added to food for farmed salmon to give the cooked salmon flesh its distinctive pink color. In Europe, astaxanthin is synthesized from petrochemicals, but synthetic pigments are not permitted in aquaculture feeds in the United States, where it is produced naturally.

← Designed for the commercial cultivation of microalgae in factories, photobioreactors are systems of well-lit narrow glass tubes in which a microalgal species grows in a nutrient solution. This maximizes photosynthesis and therefore algal production.

↗ A spectacular pink bloom of the green alga *Dunaliella salina* in the hypersaline Hutt Lagoon, Western Australia, is bordered by a wide white band of salt crystallizing in the shallow water in this arid environment.

One of the major producers of glycerol and beta-carotene is the green alga *Dunaliella salina*, which grows most luxuriantly in extreme environments under hypersaline conditions (greater than 10 percent salt), at high temperatures, and a high pH—these conditions are lethal to most contaminants of the *Dunaliella* aquaculture ponds. *Dunaliella salina* uses the large organic molecules of glycerol to maintain its osmotic balance in the hypersaline conditions. Beta-carotene is a widely occurring yellow to orange pigment found in fruits and vegetables. It is also an accessory photosynthetic pigment of the land plants, cyanobacteria, and algae. In *Dunaliella salina*, the large quantities of beta-carotene accumulate in the chloroplast, where it acts as a sunscreen. Beta-carotene protects the chloroplast from damage from the excessive light intensity by preventing the formation of reactive oxygen species. *Dunaliella* is farmed in large outdoor ponds in coastal desert regions of Israel and Australia, and in Hawaii for the beta-carotene market, which supplies the food, cosmetic, and pharmaceutical industries with a colorant, antioxidant, and an anticancer substance. Beta-carotene is the coloring agent for naturally colored margarine.

Phycocolloids

Exterior to the plasma membrane, the cell walls of red and brown seaweeds are complex structures composed of carbohydrates. Cell wall carbohydrates are distinguished by their biological function, the structurally rigid cellulose, and the flexible sulfated carbohydrates that form the matrix in which the cellulose fibers are embedded. Among many commercial applications, sulfated carbohydrates are widely used as phycocolloids, which form gels and viscous solutions in water.

In the cell walls of the red algae, the sulfated carbohydrates are either carrageenans, which are extracted from species of *Kappaphycus*, *Euchema*, *Chondrus*, *Gigartina*, and *Iridaea*; or agar, which is extracted from species of *Gracilaria*, *Agarophyton*, and *Gelidium*. Alginic acid is the principal structural carbohydrate in the brown algae, with the main sources for commercial extraction being the kelps (*Lessonia*, *Macrocystis*, and *Laminaria*) and the large brown alga *Durvillaea*.

The most cultivated seaweeds in the world are the tropical red seaweeds *Kappaphycus alvarezii* (see page 272) and species of *Eucheuma*, which together account for almost 41 percent of the 33 million tons (30 million tonnes) of seaweeds that are aquacultured annually.

← *Gelidium corneum* is harvested by scuba divers from its natural habitat at depths of 23–46 ft (7–14 m) on the northern Spanish coast.

↗ The seaweeds *Ulva* and *Gracilaria* are farmed in Israel in land-based ponds that receive flow-through seawater from the marine environment.

Over the last 40 years, these species have been introduced deliberately into 30 tropical and subtropical countries, and farmed commercially for their phycocolloids in major farms in Indonesia, the Philippines, and Malaysia.

Natural harvests of *Gelidium* along the coasts of Japan, Korea, Spain, and Portugal have recently collapsed. This, together with the fact that the large-scale aquaculture of *Gelidium* is not feasible, has resulted in a worldwide shortage of technical agar, which has serious consequences for public health.

Alginates, carrageenans, and agar are large molecules that, like gelatine, which is extracted from animals, form gels when mixed with water. Unlike gelatine, which is a gel only while refrigerated, the algal molecules gel at room temperature. They are used extensively in industry as stabilizers and binders to prevent the separation of the water and fat-soluble components of solutions. All three are used in the food industry to thicken puddings, ice cream, yoghurt, flavored milk, and fruit gels, and in the processing of canned meat, fish, and sausages. Carrageenans and alginates are also used in cosmetics and in coatings such as paints and inks. Agar is used in microbiology as the substrate for growing bacteria.

Pharmaceuticals

With tens of thousands of species in different algal lineages, the algae represent a major reservoir of bioactive compounds, which have multiple applications. Many algal species live in complex habitats, exposed to challenging conditions. In adapting to these environments, their metabolic pathways produce a wide variety of secondary, biologically active metabolites, including many that are not found in any other organisms.

BIOACTIVE CHEMICALS

Numerous algal bioactive compounds have health benefits. They protect the human body from cancer, viruses, bacteria, inflammation, diabetes, and many other health conditions. Some algae synthesize sunscreens that prevent damage to their thalli from sunlight. One compound, scytonemin—a yellow-brown pigment from the extracellular sheath of some terrestrial cyanobacteria—is the active agent added to some manufactured sunscreens. The photosynthetic phycobilin pigments of cyanobacteria and red algae protect the liver, reduce inflammation, and are antioxidants. They are also used as blue and red pigments in cosmetics.

Fucoidan, a cell wall carbohydrate of brown seaweeds, reduces inflammation and blood clotting, and has anticancer properties. One of the active agents in extracts of the brown seaweed *Sargassum muticum*, fucoidan has proved effective as an anticancer drug in two breast cancer cell lines, inducing apoptosis (programmed cell death that kills the cancer cells), and inhibiting the formation of blood vessels servicing tumors. Similar effects on lung cancer in mice have been demonstrated in extracts from the brown seaweed *Fucus evanescens*. These extracts suppressed the growth of tumors and prevented the spread of cancer cells.

REACTIVE OXYGEN SPECIES

It is generally believed that food, medicines, and cosmetics that are rich in antioxidants will neutralize reactive oxygen species, which are produced by imbalances in metabolic pathways. If the reactive oxygen species are not neutralized, they can react with nontarget molecules and cause a variety of detrimental cellular impacts, including increased cell proliferation, damage to the mitochondria and DNA, deleterious chemical chain reactions leading to the breakdown of lipids and proteins, and the inhibition of enzymes. These detrimental cellular changes may lead to cancer, cardiovascular disease, hypertension, diabetes, and the neurodegenerative Alzheimer's and Parkinson's diseases. Various algal species are now known to contain essential neutralizing antioxidants, including carotenoid (yellow and red) pigments, phenolic (tanninlike) compounds, phycobilin pigments, sulfated carbohydrates, and vitamins A and C.

← The elegant red seaweed *Asparagopsis taxiformis* is a rich source of unique biologically active chemical compounds.

↑ The common Japanese brown seaweed, and invader of temperate coasts worldwide, *Sargassum muticum* contains bioactive compounds that are effective in inducing the death of breast cancer cells.

COSMECEUTICALS

As well as their medicinal uses, bioactive compounds extracted from algae offer multiple cosmetic benefits. Bioactive compounds in algal sunscreens are added to cosmetics to protect against harmful UV light and to inhibit the formation of dark spots on the skin. Various pigments (fucoxanthin, the brown pigment of brown seaweeds, the red astaxanthin, and blue phycobilins), phenolic compounds, and fucoidans can be added to cosmetics for their anti-inflammatory, antioxidant, and anti-aging properties.

NUTRACEUTICALS

The health benefits of algae are now being recognized with their classification as "nutraceuticals"—functional foods with dietary benefits beyond their protein, lipid, and carbohydrate content. This is a crucial development that could help to counteract the increasing occurrence of dietary and lifestyle-related diseases, including type 2 diabetes, obesity, cancer, and metabolic syndrome.

↑ Powerful antioxidants found in spirulina face masks and astaxanthin capsules are claimed to have anti-aging, anti-inflammatory, and skin-whitening properties.

← Extracts with antioxidant properties from the tropical brown seaweed *Turbinaria ornata* are added to cosmetics to reduce skin aging and disorders.

↗ The Japanese kelp wakame is being trialed with other seaweeds to improve the nutritional value of processed meat products.

Traditionally, burgers, frankfurters, salami, and deli-sliced meats are high in saturated fats, salt for preservation and flavor, artificial preservatives, sodium nitrate, and the artificial flavor enhancer monosodium glutamate (the latter two chemical compounds having been linked to the occurrence of cancer). However, research is underway to evaluate the benefits of adding dried seaweed to these meat products—pork frankfurters, beef burgers, and reconstituted poultry steaks containing 5.6 percent dried milled seaweed have already passed consumer acceptance sensory tests.

Two large brown algae, wakame (*Undaria pinnatifida*) and thongweed (*Himanthalia elongata*), and the red alga laver (*Porphyra umbilicalis*), have been trialed, and the seaweed-invested meats have been shown to possess more beneficial polyunsaturated fatty acids, less salt, and more calcium, magnesium, manganese, Vitamin K, and polyphenolic compounds. When tested on animals, these meats improved lipoprotein metabolism, which suggests they could also be beneficial in humans with high blood lipid levels.

Biofuels

As crude oil is derived in part from phytoplankton that lived millions of years ago, it seems logical to explore the potential of producing biofuels from algae. Modern biofuel production requires large amounts of algal biomass, which can be derived from both microalgae and macroalgae. These can be grown in aquaculture facilities and, in the case of macroalgae, also collected from the wild.

BIOFUEL PRODUCTION

Microalgae have a high photosynthetic efficiency and high oil content, and can be grown on land that is unsuitable for food production, either in large outdoor ponds or in closed photobioreactors. Open pond systems are cheaper to operate, but poor light penetration limits photosynthesis and inefficient carbon dioxide availability can lead to low biomass productivity. Photobioreactors, which are a system of tubes, flat panels, and/or columns that expose microalgae circulating in an aqueous growth medium to light, solve these problems, but are costly to run and maintain. There are also problems associated with large-scale production.

The unicellular green algae *Chlorella* and *Scenedesmus* are two organisms that are currently cultured for biofuels, although another candidate is the freshwater green filamentous alga *Oedogonium*. This fast-growing, highly productive alga has a broad ecological distribution and builds up a large biomass in tropical and temperate freshwater habitats worldwide,

all advantageous attributes for biofuel production. It is also suitable for growing in outdoor ponds that are used for wastewater treatment.

Sugar kelp (*Saccharina latissima*) is also being investigated for efficient bioethanol production, using laminaran (the kelp's storage carbohydrate) as the fermentation substrate. Sugar kelp is a biennial or perennial species, which can grow to 13 ft (4 m) in length. Its rapid, early-season growth rates can be exploited commercially, but the demand for biomass for biofuel production will exceed the limited natural harvests of sugar kelp, requiring additional biomass to be supplied through aquaculture. Trials have already begun on establishing a sugar kelp aquaculture industry in rural eastern Maine in the United States, to both diversify local industries and preserve the traditional fishing lifestyle.

Converting seaweed into biocrude oil is an innovative approach to the disposal of stranded seaweeds on beaches. Every summer since 2011, Mexican beaches have been fouled by huge amounts of *Sargassum* from the Sargasso Sea (see pages 118, 212). The stranded seaweed causes havoc on coralgal reefs, in turtle nesting sites, and for tourism. The biomass will be converted into biocrude oil in an enclosed reactor through the process of hydrothermal liquefaction that uses moderate temperatures (350–700 °F/180–370 °C) and high pressures (100–125 bar).

↑ Biofuel is an economic solution to the *Sargassum* drifts that foul Mexican beaches.

← Sugar kelp is grown in Maine in the United States to produce bioethanol.

Toxic algae

Phytoplankton are important primary producers in aquatic environments. On most occasions, phytoplankton blooms of 4.5 million cells per gallon (1 million cells per liter) are beneficial to fisheries production and aquaculture. Sometimes the rapid growth rates, often of one dominant microscopic species, generate sufficiently high cell numbers to discolor the water (usually red in the sea), causing what is known as a "red tide" or "harmful algal bloom."

TOXIC SPECIES

Globally, toxic cyanobacterial blooms occur commonly in degraded freshwater lakes, drinking-water reservoirs, estuaries, and bays, while dinoflagellates are the most common bloom-forming species in the marine environment. Dinoflagellates produce some of the most potent toxins on Earth and are responsible for more than 60,000 poisonings per year. Thankfully, of the more than 5,000 marine phytoplankton species, only a small proportion (roughly 300 species) form red tides, and an even smaller proportion (at least 40 to 60 species) are known to be toxic. This latter group includes some species of dinoflagellates, and, to a much lesser extent, diatoms.

During "red tide" blooms, very high concentrations of dinoflagellate cells may impart spectacular pink or red discolorations of the sea surface in some coastal areas worldwide. On the northeast coast of New Zealand, spring blooms are often dominated by the dinoflagellate *Noctiluca scintillans*, a nontoxic species well known for producing tomato-red blooms. However, in areas where mass mortalities of tens of thousands of fish (for example, flounder, mullet, eel, and goby) and of 8,500 farmed abalone were observed, several dinoflagellate species, including three species of *Karenia* and *Noctiluca scintillans*, bloomed simultaneously.

During one of these New Zealand blooms, a maximum of 150 million cells per gallon (33 million cells per liter) of *Karenia* species turned the sea surface red. Toxic *Karenia* species are known to kill a wide range of marine life. Shellfish (mussels, pipis, cockles, and oysters), which filter feed on phytoplankton, accumulate neurotoxins (brevetoxins) from toxic *Karenia* species in their tissues, which, if eaten by humans, cause illness, respiratory distress, and sometimes death. It is thought that the prevailing El Niño conditions may have been responsible for the increased number and intensity of these blooms.

Some benthic dinoflagellate species are toxic. The usually tropical seaweed-dwelling species of *Gambierdiscus* are responsible for the human illness ciguatera fish poisoning, which affects around 50,000 people annually. Ciguatoxins bioacummulate in food webs when smaller herbivorous fish are preyed on by carnivorous fish and reach their highest concentration in the large fish, which may be eaten by humans.

→ An aerial image of a huge, spectacular "red tide" bloom caused by dinoflagellates at Leigh, near Cape Rodney, on the northeast coast of New Zealand.

ALGAL TOXINS

Many algal toxins have been identified, and more are being discovered—often after the poisoning of humans and animals. Saxitoxins are potent neurotoxins that are produced by marine and freshwater organisms spanning two kingdoms of life. Several species of marine dinoflagellate genera (*Alexandrium*, *Pyrodinium*, and *Gymnodinium*) are responsible for around 2,000 cases of human saxitoxin poisoning every year, while saxitoxin-producing species of freshwater cyanobacteria (*Dolichospermum*, *Aphanizomenon*, and *Planktothrix*) are considered a serious toxicological risk to drinking water. Humans can ingest saxitoxins either by drinking water that has been contaminated by toxic cyanobacterial blooms, or by eating contaminated shellfish that have been filter feeding on toxic dinoflagellates, which leads to paralytic shellfish poisoning (PSP). Paralytic shellfish poisoning in humans causes numbness on the face and neck, headache, dizziness, nausea, vomiting, and diarrhea. In extreme cases, it can progress to muscular paralysis, pronounced respiratory difficulties, a choking sensation, and finally death.

Another shellfish poisoning produces nausea, vomiting, severe diarrhea, and stomach cramps in humans. First documented in Japan in 1974, the shellfish that caused diarrheic shellfish poisoning were found to be contaminated with okadaic acid, which is a toxin produced by the dinoflagellate *Dinophysis fortii*. Since then, several species of *Dinophysis* (see page 226) have been identified as producing okadaic acid and dinophysistoxins.

← During algal blooms, algal toxins and/or low dissolved oxygen levels can cause massive fish kills.

↗ A light microscope image of a large coiled filament of *Dolichospermum circinale* (formerly *Anabaena circinalis*), a common toxic cyanobacterium notorious for causing freshwater algal blooms worldwide.

In 1995, several cases of human poisoning resulting from the consumption of aquacultured blue mussels were reported in the Netherlands. The symptoms were typical for diarrheic shellfish poisoning syndrome, but chemical analysis of the contaminated mussels revealed the presence of low levels of okadaic acid and dinophysistoxins, which indicated that *Dinophysis* was not responsible.

The causative toxin was later identified as the novel biotoxin azaspiracid, although pinpointing the culprit was complicated. Initially, the dinoflagellate *Protoperidinium*

crassipes was thought to be the cause, but while cells collected from the sea contained the azaspiracid toxin, laboratory cultures of the same species did not produce azaspiracid. Fortuitously, in 2007—more than a decade after the initial poisoning event—scientists on a research vessel in the North Sea detected azaspiracid in the seawater. This allowed phycologists to hone in on the causative agent, which was discovered to be the new dinoflagellate genus and species *Azadinium spinosum*. It was by eating these dinoflagellates that *Protoperidinium crassipes* had acquired the azaspiracid toxin.

Climate change

The level of atmospheric carbon dioxide has been increasing since the Industrial Revolution. This has led to rising air and sea surface temperatures followed by climate change. Higher atmospheric carbon dioxide levels also result in more carbon dioxide dissolving in seawater, which makes the oceans more acidic; the pH of seawater has reduced by 0.12 units compared to preindustrial levels. Both higher temperatures and ocean acidification will potentially affect algal species.

SEA SURFACE TEMPERATURES

Temperature profoundly affects the recruitment, growth, survival, and reproduction of algal species, and is a major determinant of algal distribution patterns worldwide. Many algal species live close to their upper lethal temperature limits, at the trailing edge of their geographical ranges. For these macroalgal species an increase in sea surface temperatures can lead to a shift in their geographical range and local extinctions of marginal populations. Even small increases in temperature are enough to impact these marginal populations, leading to a poleward shift in their geographical range.

Kelps that are largely restricted in their geographical distribution to cool and cold temperate coasts are especially sensitive to rising sea surface temperatures. A worldwide decline in submarine kelp forests has already been reported, and this is considered to be at least partially due to an increase in sea surface temperatures. El Niño Southern Oscillation (ENSO) events are a complex interaction between the atmosphere and the Pacific Ocean that compound the effects of climate change. In the southern hemisphere, the southward-flowing warm East Australian Current flows more strongly during ENSO events, transporting warm, nutrient-poor water to the cool temperate Tasmanian coast, which decimates the *Macrocystis* forests. Californian kelps also die during ENSO events, when the usual upwellings of cold nutrient-rich water fail.

OCEAN ACIDIFICATION

Acidification of the ocean also has the potential to affect algal species, as it alters the carbonate chemistry of the typically alkaline seawater. When excess atmospheric carbon dioxide dissolves in seawater it shifts the carbonate buffer system in seawater to a bicarbonate-dominated system. This makes calcification more difficult for calcifying algae, and their calcium carbonate skeletons will be less dense.

↑   Geographical ranges of the intertidal bladder, knotted, and serrated wracks have shifted poleward with increasing temperatures.

←   Kelp forests lose many individual kelp thalli in response to a warming climate.

Ocean acidification threatens the survival of the large populations of the calcified green algae *Halimeda* and its calcified relatives, as well as the coralline red algae that are the foundation species and ecosystem engineers on coralgal reefs and rhodolith communities.

Coralline red algae are considered particularly susceptible to ocean acidification, as they are constructed of both calcium carbonate-calcite and magnesium carbonate-calcite, the latter the most soluble form of calcium carbonate. The encrusting coralline red alga *Crusticorallina muricata* (see page 108), growing at Tatoosh Island, Washington State, on the west American coast is already showing signs of ocean acidification. Crust samples of this species taken close to the growing edge of the thallus in 2012 had a thickness that was just half that of comparable samples collected between 1981 and 1997.

KINGDOM	: Chromista
PHYLUM	: Haptophyta
CLASS	: Coccolithophyceae
ORDER	: Isochrysidales
GENUS	: *Emiliania*
SIZE	: Cell diameter 5–10 microns
HABITAT	: Pelagic

Emiliania huxleyi

Coccolithophorid

Although more than 200 species of haptophytes have been described, *Emiliania huxleyi* dominates the phytoplankton in subpolar regions and also occurs in the deeper waters of subtropical seas.

Emiliania huxleyi frequently forms large monospecific blooms—some covering more than 95,000 sq miles (250,000 sq km), with cell densities of up to 45 million cells per gallon (10 million cells per liter) of seawater. Although the cells are small, the calcite coccoliths on the cell surface, which can become detached and float in the water column, are highly reflective, enabling these blooms to be seen from aircraft and satellites.

This ecologically versatile species is widely distributed in the marine environment. Its bloom-forming capacity in both coastal and oceanic regions is achieved by diversifying into different strains. Coastal strains are more tolerant of lower salinities than oceanic strains, and warm water strains are genetically adjusted to higher temperatures than temperate strains. The ecological success of this species is due in part to the competitive advantages conferred by their rapid uptake of nutrients after the spring diatom blooms, their adaptation to high light intensities at the sea surface, and high growth rates. Compared to other coccolithophorids, some strains of *Emiliania huxleyi* are extremely fast growing.

CARBON CYCLE

Emiliania huxleyi is a major contributor to the carbon cycle, as it deposits calcite in its coccoliths equal to the organic content of the cells. When the cells die, the heavy calcite coccoliths sink to the deep seafloor where they accumulate as inorganic carbon (carbonate) deposits. These deposits, which lock carbon away from the atmosphere for millions of years, are quantitatively important—they cover one-half of the seafloor which is equivalent to one-third of the Earth's surface.

SULFUR CYCLE

Emiliania huxleyi also plays a central role in the sulfur cycle, by releasing high concentrations of the volatile dimethyl sulfide into seawater. Coccoliths of the living cells reflect light away from the Earth's surface and dimethyl sulphide enhances cloud formation further cooling the planet. This species, unknown to many people, plays an enormous role in cooling Earth's climate.

→ This electron microscope image of the cells of *Emiliania huxleyi* reveals a cell covering of many calcareous circular scales (coccoliths). The coccolith's upper plate is reminiscent of a spoked wheel, and the lower plate a disk (see bottom left of image).

PHYLUM RHODOPHYTA

Neopyropia yezoensis

Nori

KINGDOM	Plantae
PHYLUM	Rhodophyta
CLASS	Bangiophyceae
ORDER	Bangiales
GENUS	*Neopyropia*
SIZE	Thallus to length of 12 in (30 cm)
HABITAT	Open rocky coasts

The most famous seaweed in the world, nori is used as the wrap around sushi rolls. The Japanese have been making the papery nori wrap since the 1600s, when they applied their newly invented papermaking techniques to seaweed.

The process involved taking the harvested nori, shredding it into a slurry, pressing it into paper-thin sheets in a frame, and drying it in the sun. Nowadays, nori papermaking is fully automated. From the mid-1600s, *Neopyropia tenera* (traditionally known as Asakusa-nori) was farmed in Tokyo's Asakusa district, and this tradition continued for three centuries until coastal development around Tokyo Bay in the mid-twentieth century destroyed nori's estuarine habitat. Today, though, the most widely cultivated nori species in Japan is *Neopyropia yezoensis*, which is known to the Japanese as *Susabi-nori*.

Nori's nutritional value lies in its high protein levels (30–50 percent), mineral composition, and extremely high vitamin content (A, thiamine, B_2, niacin, B_6, and C). The best-tasting nori contains high levels of the flavor-inducing amino acids taurine, glutamic acid, asparagine, and alanine, combined with a subtle sweetness from the free sugars, particularly isofloridoside.

Nori is a fall-to-winter crop, which is farmed in estuaries and sheltered bays, on nets either strung between poles or, in deeper water, attached to buoys anchored to the seafloor. Each September over recent decades, the nori fishermen tied small mesh bags at regular intervals along the long nets. The bags, which contained several oyster shells covered with pink spots—the filamentous stage in the nori life history—released spores that attached to the nets and grew into nori blades. The first harvest in December produces the most flavorsome blades, but there will be several harvests after that.

Approximately half of Japan's nori crop is produced in the Ariake Sea, a very large enclosed bay on Kyushu, the country's southernmost main island. Japan's highest-quality nori is grown on nets strung between poles in Saga Prefecture at the northern end of the Ariake Sea.

→ The most flavorsome Japanese nori is grown on horizontal grid nets, 59 ft (18 m) long and 5 ft (1.5 m) wide with a 12 in (30 cm) stretched mesh, strung between poles. To maximize flavor, it is exposed to the atmosphere during low tides.

KINGDOM	:	Chromista
PHYLUM	:	Ochrophyta
CLASS	:	Phaeophyceae
ORDER	:	Laminariales
GENUS	:	*Saccharina*
SIZE	:	Thallus length 6–11.5 ft (1.8–3.5 m)
HABITAT	:	Subtidal zone on open coasts

PHYLUM OCHROPHYTA

Saccharina japonica

Ma konbu

Twelve species of the kelp genus *Saccharina* are the foundation species that build the submarine kelp forests of northern Japan.

Konbu kelps inhabit the cold temperate coasts of Hokkaido—the northernmost of Japan's four main islands. Commonly known as "konbu" (pronounced "kombu") in Japan, nine of the species are consumed by the Japanese, including the highly prized ma konbu (*Saccharina japonica*). Ma konbu generally grows from 6.5 to 16.5 ft (2–4 m) in length and has a simple kelp morphology, comprising a holdfast, stipe, and an undivided leathery blade.

Ma konbu is aquacultured and harvested from the wild. The wild harvesting removes the attached kelps from rocks and detached drifting kelps from the shore. Far more of the seaweed is farmed, the fishermen using horizontal, gridlike, raft and rope systems anchored to the seafloor and buoyed at a depth of 6.5–23 ft (2–7 m) below the surface. In 2016, just over 64,000 tons (58,111 tonnes) of ma konbu was wild-harvested, compared to more than 9 million tons (8.2 million tonnes) farmed in coastal waters around Hokkaido. Konbu is valued for its micronutrients, particularly iodine, which is essential for the production of thyroid hormones and for its other health benefits. It is known to reduce hypertension (high blood pressure) and blood cholesterol levels, while its high levels of the carbohydrate fucoidan are thought to be important in fighting cancer and slowing aging.

TASTE SENSATION

In 1908, Dr. Kikunaye Ikeda discovered that glutamic acid was the taste sensation that made konbu popular with the Japanese, and this paved the way for the production of monosodium glutamate. Today, konbu is indispensable in Japanese cuisine. It is not only one of the main ingredients in dashi stock, but is also sold dried and eaten after softening with rice vinegar; dried, shaved, and served as tororo konbu; and stewed in a strong soy sauce to make konbu tsukudani. The famous Osaka dish battera sushi is made with konbu, mackerel, and vinegared rice.

→ In northern Japan, the long, leathery blades of konbu are preserved in summer by being spread out on the beach and dried in the sun.

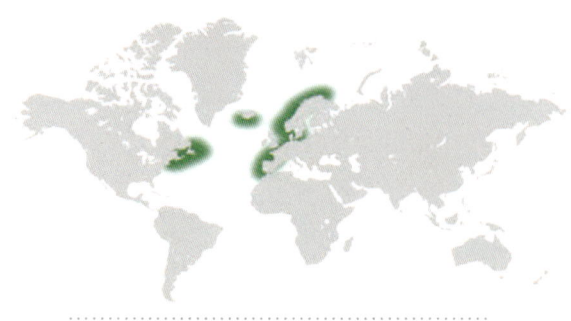

Chondrus crispus

Irish moss, carrageen

KINGDOM	Plantae
PHYLUM	Rhodophyta
CLASS	Florideophyceae
ORDER	Gigartinales
GENUS	*Chondrus*
SIZE	Thalli rarely exceed 8 in (20 cm) in length
HABITAT	Mid to lower rocky seashores to depths of 65 ft (20 m)

The Irish moss (*Chondrus crispus*) is a dark, reddish-purple seaweed with thin fronds that broaden out above the holdfast into branching fans. This common red seaweed grows as a dense covering on rocks, and is a conspicuous component of lower intertidal algal communities on both sides of the cool temperate North Atlantic Ocean.

The Irish moss has long been known for its gelling properties, with the first commercial harvests dating back to the late nineteenth century in the coastal town of Carragheen, Ireland (from where the phycocolloid, carrageenan, derives its name). In North America, commercially harvested carrageenan was used mainly in home cooking and in cough syrups, before an extraction plant was established in New England in the 1930s to supply the stabilizer for chocolate milk.

COMMERCIAL USES

Following a shortage of agar from Japan during World War II, the West continued to explore commercial uses of carrageenan, and it is now used widely in the food industry as a thickener and stabilizer, especially for milk products, ice cream, and processed foods, such as canned meats, fruit gels for babies, instant desserts, and salad dressings. Carrageenan is sometimes referred to as "vegetarian gelatin"—a plant-based substitute for the animal-derived gelatin. It is used as the gelling agent in the traditional Irish moss pudding, which is a sweet, silky pudding with a very soft jelly set that is much softer than panna cotta. Currently, most carrageenan is produced by farming the tropical red seaweeds *Eucheuma* and *Kappaphycus alvarezii*.

As well as its uses in the food industry, carrageenan is also used widely in pharmaceuticals, cosmetics, and in coatings such as paints and inks. However, perhaps its most well-known application is in traditional cough mixtures that are sold around the world. Medicinally, Irish moss cough mixtures are thought to have antiviral and expectorant (phlegm-loosening) properties that can help clear chest infections and ease coughs.

→ Traditionally eaten by humans for centuries, the red seaweed *Chondrus crispus* is still foraged from the seashore and eaten by the Irish. Its common name of Irish moss is misleading, since it is not actually a moss.

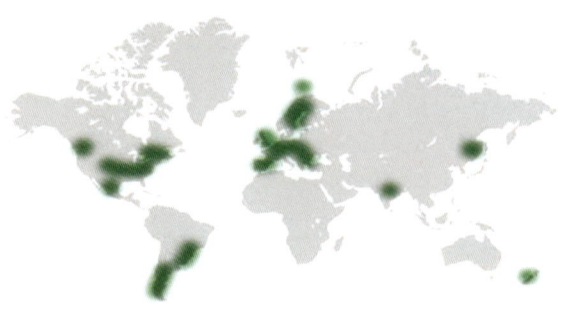

PHYLUM CHLOROPHYTA

Haematococcus lacustris

Blood-red alga

KINGDOM	Plantae
PHYLUM	Chlorophyta
CLASS	Chlorophyceae
ORDER	Chlamydomonadales
GENUS	*Haematococcus*
SIZE	Cells 15–50 microns long
HABITAT	Small freshwater rock pools, ponds, and ornamental birdbaths

The algae synthesize a wide variety of functionally important pigments that are also a source of bioactive chemicals for human use. The reddish to yellow carotenoid pigments assist with photosynthesis but they are also found in the cytoplasm of algal cells.

Beta-carotene, which colors carrots, is widely distributed in algal cells, while red carotenoid pigment, astaxanthin, is produced by relatively few organisms, including some marine bacteria, a red yeast, and a few green algal species, most notably the freshwater unicellular green microalga *Haematococcus lacustris*.

Astaxanthin is a high-value carotenoid, which has been dubbed "super vitamin E." Compared to other natural antioxidants, such as vitamins C and E, and beta-carotene, astaxanthin has a greater capacity to scavenge the damaging reactive oxygen species. This potent biological antioxidant is therefore considered an effective agent for the prevention of age-related, degenerative, and chronic diseases, such as cataracts, macular degeneration, and atherosclerosis, as well as some types of cancer. Astaxanthin is also used in food dyes and feed additives in aquaculture and poultry farming, particularly to color the flesh of aquacultured salmon and trout pink.

Of all the astaxanthin-producing organisms, *Haematococcus lacustris* is the richest reported biological source of the pigment, containing up to 4 percent dry weight, although its slow growth rate and culture difficulties currently restrict its application on a large scale. This blood-red algal species accumulates the carotenoid in high concentrations when its cells are subjected to the stress conditions of nutrient starvation, strong sunlight, increasing salinity, and high or low temperatures. It is the accumulation of astaxanthin in lipid vacuoles in the cytoplasm that causes the microalga's cell color to change from green to red.

A common inhabitant of birdbaths, the blood-red algae are well adapted to the drying up of their habitat. In adverse conditions, the cells synthesize and accumulate the red pigment and form a thick-walled dormant stage. The thick walls allow these algae to survive the desiccation associated with the drying out of the birdbath while the pigment— a sunscreen—protects the chloroplast from the damaging sunlight and ultraviolet light.

→ This light microscope image of the unicellular green alga commonly known as the blood-red alga shows the red pigment synthesized under stress conditions to act as a sunscreen in the cytoplasm of its cells.

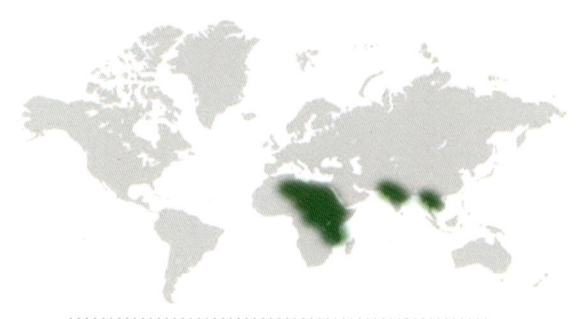

KINGDOM	Eubacteria
PHYLUM	Cyanobacteria
CLASS	Cyanophyceae
ORDER	Oscillatoriales
GENUS	*Limnospira*
SIZE	Filaments up to 0.08 in (2 mm) or more in length, 8–12 microns in diameter
HABITAT	Soda (alkaline and saline) inland lakes

PHYLUM CYANOBACTERIA

Limnospira fusiformis

Spirulina

Species of *Limnospira* have regularly coiled filaments, which are composed of cells that are clearly demarcated by visible cross walls (septa).

Three cyanobacterial genera—*Arthrospira*, *Spirulina*, and the newly described *Limnospira*—have helically coiled filaments that appear superficially similar, which explains why *Limnospira fusiformis* was known for a long time as *Spirulina fusiformis,* and more recently as *Arthrospira fusiformis,* and why it is sold in the health food market as "spirulina." However, recent DNA sequencing studies have revealed that these three genera are distinct entities and are not closely related genetically.

Limnospira fusiformis blooms in shallow brackish lakes, particularly the soda lakes in the African Rift Valley. These inland lakes are both saline and alkaline, with the alkalinity (pH 10) caused by hot springs delivering sodium carbonate-rich water into the lakes (hence the name "soda lakes"). *Limnospira fusiformis* and *Limnospira maxima* are both produced commercially for the health food market in large outdoor ponds that are either open to the atmosphere or covered with transparent material. A third species, *Spirulina platensis*, is known scientifically with certainty from a collection long ago from saline lakes in Uruguay, South America, although the name has been used incorrectly for decades as the source of the "spirulina" powder.

FOOD AND VITAMIN SOURCE

Spirulina has a long history as a food source. The Spanish described Aztec fishermen collecting spirulina from Lake Texcoco in Mexico, which they prepared as a dry cake. Women living around Lake Chad in Africa have, for centuries, harvested and used spirulina in meat and vegetable broths. "Spirulina" powder has a high nutritional value based on its high protein content (up to 70 percent), vitamins, essential polyunsaturated fatty acids, pigments, and antioxidants.

Spirulina also has a very high vitamin B_{12} content. This vitamin is mainly found in foods derived from animals. Its presence in large quantities in "spirulina" powder is particularly beneficial for plant-based diets. In fact, just 0.7 ounces (20 g) of spirulina will provide an adult with their daily requirement of vitamin B_{12} as well as 70 percent of their daily requirement of vitamin B_1, and 50 percent of vitamin B_2. Spirulina extracts also exhibit antiviral, antimicrobial, antioxidant, anti-inflammatory, and antidiabetic properties.

→ The microscopic coiled filaments of spirulina, a popular health food, are a rich source of protein and B-complex vitamins that are also acceptable to vegetarians.

KINGDOM	Plantae
PHYLUM	Rhodophyta
CLASS	Florideophyceae
ORDER	Gigartinales
GENUS	*Kappaphycus*
SIZE	Thallus to 6.5 ft (2 m) long
HABITAT	Attached to rubble on coralgal reefs, from the intertidal zone to a depth of 65 ft (20 m)

Kappaphycus alvarezii

Spiny farmed alga

Kappaphycus alvarezii is among the larger red algae, with thalli that can grow to 6.5 ft (2 m) in length and weigh in at 55 pounds (25 kg). The thallus is composed of multiple tough, fleshy, firm main axes and branches, which can reach 0.4–0.8 in (1–2 cm) in diameter and are covered with irregularly arranged blunt or spiny protuberances.

Farming species of *Euchema* began in the Philippines in the early 1970s, based on research undertaken by Professor Maxwell Doty, a phycologist working at the University of Hawaii. In the late 1970s, Doty discovered the fast-growing species *Euchema alvarezii* growing on a coral reef in Sabah, Malaysia, which subsequently was reclassified as *Kappaphycus alvarezii*. This species is one of the most important sources for the extraction of carrageenans.

This early enterprise proved so successful that farming these red algae for carrageenans spread to more than 30 countries in tropical southeast Asia and east Africa. *Kappaphycus alvarezii* appears to be have been deliberately introduced to many of these countries from its original source in Sabah, Malaysia. Today, the Philippines, Indonesia, and Malaysia are the major producers of carrageenans,

with lesser production in Vietnam, Cambodia, Myanmar, and Tanzania. The economic benefits of seaweed farming flow through coastal communities, and there are now tens of thousands of people—many of whom live in economically deprived coastal areas—engaged in the industry.

Farming *Kappaphycus* relies on vegetative propagation. A plant that is deemed to be healthy, a good color, and weighing a little over 2 pounds (approximately 1 kg) will yield six to eight cuttings for the next cultivation cycle. These cuttings are tied either in singlets or doublets on to a rope, which is attached to stakes driven into the bottom of the shallow sea. Mature plants are harvested 30, 45, or 60 days after the outplanting of the cuttings, depending on the site or the financial needs of the growers.

As well as its gelling properties, *Kappaphycus alvarezii* has been shown to contain chemical compounds that have antioxidant, antibacterial, antidiabetic, and anticancer properties, and there is now a lot of interest in exploring its biotechnological applications.

→ The harvesting of *Kappaphycus alvarezii* is a cottage industry in coastal communities in Asia and Africa. The large plants are dried in the tropical sun by the growers before being sent to factories for phycocolloid extraction.

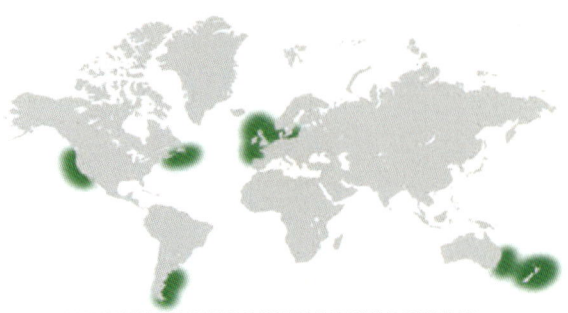

KINGDOM	Chromista
PHYLUM	Bacillariophyta
CLASS	Bacillariophyceae
ORDER	Bacillariales
GENUS	*Pseudo-nitzschia*
SIZE	Cells 90–130 microns long and 4–5 microns wide
HABITAT	Coastal plankton

PHYLUM BACILLARIOPHYTA

Pseudo-nitzschia multiseries

Toxic diatom

The long narrow cells of the pennate diatom *Pseudo-nitzschia multiseries*, which measure 90–130 microns long but only 4–5 microns wide, have to be viewed in an electron microscope to distinguish this species from the other *Pseudo-nitzschia* species.

Species of *Pseudo-nitzschia* form long, stepped colonies of 10 or more cells, the tips of the neighboring cells overlapping to roughly one-third of the cell's length.

In November 1987, 100 people unexpectedly became ill and three people died of a mysterious new syndrome linked to the consumption of mussels from eastern Prince Edward Island, Canada. The new syndrome—amnesic shellfish poisoning—produced symptoms of dizziness, nausea, vomiting, amnesia, and death in humans, and it is now known to cause similar symptoms in marine mammals and birds.

The toxin was quickly identified as the neurotoxic amino acid, domoic acid. Mussels filter feed on phytoplankton, a fact which implicated a phytoplankton species as the causative agent of the poisonings. However, it took several months to identify the culprit, which turned out to be a pennate diatom. This was surprising, as it was the first time that a diatom had been identified as a toxin producer—until then, most poisonings from algal blooms had been caused by dinoflagellates and cyanobacteria.

The toxin producer was initially identified as *Nitzschia pungens forma multiseries*, which had bloomed around Prince Edward Island at the time of the poisonings; blooms are generally necessary if shellfish are to accumulate enough toxic phytoplankton to poison humans. The genus *Nitzschia* is the second largest diatom genus, comprising approximately 900 species, but having gained notoriety for the poisonings, a closer examination revealed that it was not a species of *Nitzschia* after all, but *Pseudo-nitzschia multiseries*, a species of the new genus *Pseudo-nitzschia*. At least 10 species of *Pseudo-nitzschia* are now known to produce domoic acid, although not all strains of these species produce toxins.

Routine monitoring programs for the presence of algal toxins in aquacultured shellfish (mussels, oysters, clams, and other bivalves) have been instigated in coastal areas known to be inhabited by toxic algal species. If algal toxins are found in the shellfish, the farms are temporarily closed until the bloom passes and the shellfish lose their toxicity. These monitoring programs have reduced the incidence of human poisonings. In addition, signs are often erected to warn the public of the hazards of harvesting wild shellfish in areas prone to toxic algal blooms.

→ A light microscope image of a *Pseudo-nitzschia* bloom. The characteristic chains of the genus are formed by overlapping ends of (in this image, two to four) long, spindle-shaped cells, with each cell containing two yellow-brown plastids.

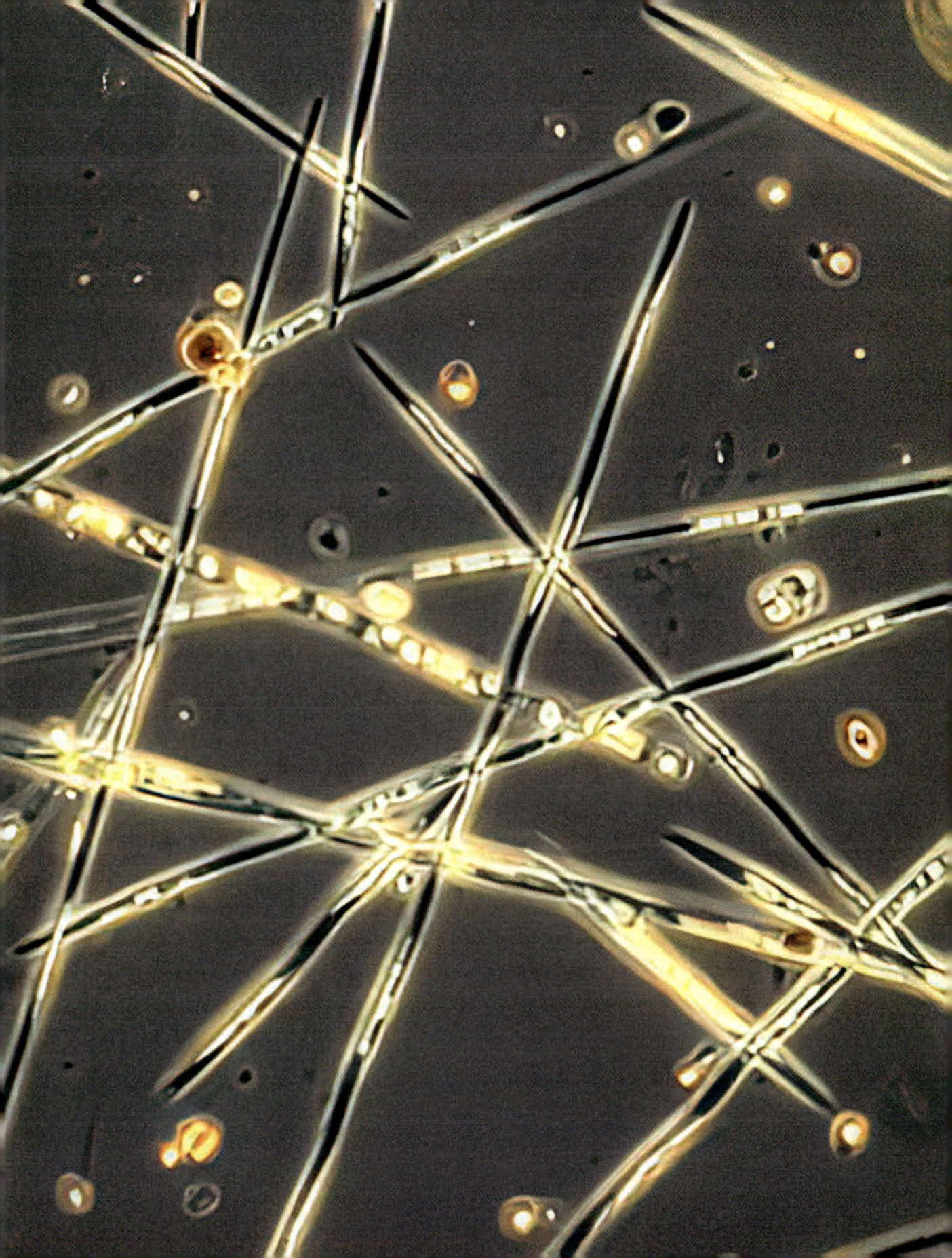

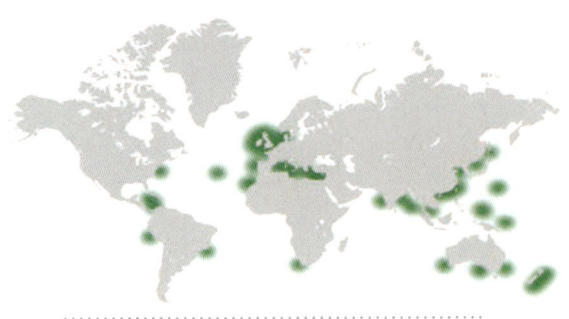

KINGDOM	:	Chromista
PHYLUM	:	Miozoa
CLASS	:	Dinophyceae
ORDER	:	Gonyaulacales
GENUS	:	*Alexandrium*
SIZE	:	Cells 22–29 microns long
HABITAT	:	Coastal plankton

CLASS DINOPHYCEAE

Alexandrium minutum

Toxic dinoflagellate

Alexandrium minutum was described in 1960 when this dinoflagellate species caused red discoloration of the water in Alexandria Harbour, Egypt. The cells of this species are small, spherical, and armored, with delicate pores on the surface of the thecal plates.

Like several other species of *Alexandrium*, *Alexandrium minutum* is known to produce paralytic shellfish toxins, neurotoxins that cause the syndrome known as "paralytic shellfish poisoning." The saxitoxins accumulate in bivalve shellfish as they filter feed during blooms of the toxic dinoflagellates.

PARALYTIC SHELLFISH POISONING

One of the earliest documented fatal cases of paralytic shellfish poisoning occurred in 1793, when five members of Captain George Vancouver's crew became ill, and one died, after eating mussels collected in an area now known as Poison Cove on the central British Columbian coast, Canada. The captain noted that it was taboo for the local First Nations people to eat shellfish when the sea became bioluminescent, which is a phenomenon that is often caused by dinoflagellate blooms.

Within 30 minutes of ingesting toxic shellfish, mild cases of paralytic shellfish poisoning present symptoms of numbness around the lips, which gradually spreads to the face and neck; as well as a headache, dizziness, nausea, vomiting, and diarrhea. In extreme cases, patients experience muscular paralysis, pronounced respiratory difficulties, and a choking sensation. Death due to respiratory paralysis may occur 2–24 hours after ingestion.

In 1927, with cases of paralytic shellfish poisoning continuing to be reported along the North American west coast, the state of California recognized that the syndrome presented a serious health risk and began a prevention program. In 1937, the dinoflagellate that is today known as *Alexandrium catenella* was identified as the causative agent for these shellfish poisonings, and to date, six toxic species of *Alexandrium* have now been identified along this coastline. Toxic algal blooms continue to be a problem on the North American west coast, with Pacific coast states monitoring for the presence of algal toxins in shellfish.

→ This electron microscope image of a cell of *Alexandrium minutum* shows the thick thecal plates (the armor) covering the cell surface, tiny pores on the thecal plates, and the transverse and longitudinal furrows. The transverse and longitudinal flagella, normally located in the furrows, have been lost from the cell during processing for the electron microscope.

CHARACTERS USED TO DEFINE
SELECTED ALGAL PHYLA AND GROUPS

PHYLUM/GROUP	MAJOR PHOTOSYNTHETIC PIGMENTS	CELL COVERING
Cyanobacteria	Chlorophyll *a*, phycobilins	Peptidoglycan cell wall
Glaucophyta (blue-gray algae)	Chlorophyll *a*, phycobilins	Flattened vesicles with plates beneath the plasma membrane
Rhodophyta (red algae)	Chlorophyll *a*, phycobilins	Cellulose cell wall with mucilaginous agar, carrageenan or porphyran
Chlorophyta (green algae)	Chlorophyll *a*, *b*	Glycoprotein membrane or organic scales, or a mainly cellulose cell wall
Charophyta (charophytes)	Chlorophyll *a*, *b*	Cellulose cell wall
Euglenozoa (euglenoids)	Chlorophyll *a*, *b*	Proteinaceous pellicle with sliding bands under the plasma membrane
Chloroarachniophyceae (green spider algae; green amoebae)	Chlorophyll *a*, *b*	Naked (only a plasma membrane)
Crytophyta (cryptophytes)	Chlorophyll *a*, *c*, phycobilins	Proteinaceous periplast often of hexagonal, rectangular, oval, or round plates
Haptophyta (haptophytes, coccolithophorids)	Chlorophyll *a*, *c*, fucoxanthin	Calcium carbonate scales, called coccoliths, common
Dinophyceae (dinoflagellates)	Chlorophyll *a*, *c*, peridinin	Vesicles under the plasma membrane often containing cellulose plates
Heterokonts (large group: 18 classes)	Chlorophyll *a*, *c*, some with fucoxanthin	Variable, some naked or with silica or organic scales or with a cell wall of cellulose, alginates, and fucoidan

PLASTID ENVELOPE	PLASTID ORGANIZATION	FLAGELLA NUMBER, FORM, AND ARRANGEMENT
Plastid lacking	Lacks plastids, single thylakoids in cytoplasm	Lack cells with flagella
Peptidoglycan wall	Single thylakoids in a primitive plastid	Two smooth, unequal flagella, subapically inserted into cell
Two membranes	Single thylakoids	Lack cells with flagella
Two membranes	Stacks of two to six to many fused thylakoids	Usually two or four (sometimes one or eight), smooth, equal flagella apically inserted into cell
Two membranes	Stacks of one to three thylakoids	Two, equal, subapically inserted into the cell
Three membranes	Stacks with three thylakoids	Usually two, inserted into an apical invagination, one long emergent hairy flagellum, one barely emergent flagellum
Four membranes, reduced nucleus between the outer and inner membrane pairs	Stacks of one to three thylakoids	One hairy flagellum subapically inserted into the cell
Four membranes, reduced nucleus between the outer and inner membrane pairs	Stacks of two thylakoids	Two, equal or subequal, with one or two rows of stiff hairs, inserted into an apical depression
Four membranes	Stacks of three thylakoids	Two, equal or sub-equal, generally smooth, apically inserted: haptonema present
Three membranes	Stacks of three thylakoids	Two, unequal, one helical with a single row of long hairs in girdle, other smooth trailing, insertion lateral
Four membranes	Stacks of three thylakoids frequently with a girdle lamella bounding the thylakoid stacks	Variable, typical heterokont flagella; two, unequal, anterior flagellum with two rows of stiff hairs; smooth posterior flagellum

GLOSSARY

agar Complex mucilaginous carbohydrate in the cell wall of some red algal species.

akinete Thick-walled asexual spore that often undergoes a period of dormancy.

alginate Complex structural carbohydrate in the cell wall of brown seaweeds.

alternation of generations A life history pattern in which the haploid (n) phase (or thallus) alternates with the diploid (2n) phase (or thallus).

alternation of heteromorphic generations A life history pattern in which the haploid and diploid phases are morphologically different.

alternation of isomorphic generations A life history pattern in which the haploid and diploid phases are morphologically similar.

anaerobic **1.** An environment in which oxygen is absent. **2.** An organism only able to survive in the absence of oxygen. **3.** An enzyme that only functions in the absence of oxygen.

anisogamy (adj. anisogamous) A type of sexual reproduction in the algae in which the flagellate male and female gametes differ in size.

apical cell The cell at the apex (tip) of the thallus that is often capable of cell division.

apical growth Growth that occurs by the division of one or more cells at the apex (tip) of the thallus.

articulated coralline red algae Red seaweeds with erect thalli composed of alternating calcified and uncalcified segments (or joints).

asexual reproduction Any reproductive process that has only one parent and does not involve the union of gametes.

benthos (adj. benthic) Organisms that live in, on, or attached to the bottom sediments of aquatic ecosystems.

binary fission A relatively simple method of cell division most common in bacteria in which cells divide into two progeny cells.

bioluminescence Production of light by living organisms.

biomass The weight of living organisms in an ecosystem at any moment in time.

blade A thin, flat, sheetlike algal thallus.

carrageenans Complex mucilaginous carbohydrates in the cell walls of some red seaweeds.

centric diatom A diatom that has radial symmetry.

charophytes (phylum Charophyta) Species of freshwater green algae most closely related to the land plants.

ciliate A motile (swimming) unicellular organism characterized by a cell covered by numerous cilia (short hairs), the presence of a larger nucleus and a smaller nucleus, and a cell mouth.

chloroarachniophytes (class Chloroarachniophyceae) Species of green spider algae classified in the kingdom Protozoa.

coccolithophorids Unicellular species of the haptophyte algae that have a cell covering of small ornate calcareous scales (coccoliths).

coccoliths Calcium carbonate scales covering the cell surface of the unicellular planktonic coccolithophorids.

coenocytic thallus A multinucleate thallus without septa (cross walls) and therefore without cells.

colony A group of cells held together by mucilage or cell wall material.

cryptophytes (phylum Cryptophyta) Species of unicellular flagellate algae, without or with plastids derived from a red alga.

diffuse growth A type of growth pattern that occurs over most of the thallus rather than in a localized region of cell division.

dinoflagellate (class Dinophyceae) Unicellular or colonial algae that possess vesicles underneath the plasma membrane, have two distinct (one transverse, one longitudinal) flagella, and may have or lack plastids.

endosymbiont A partner in a symbiosis that lives inside the cells or tissues of the other symbiont.

eukaryote An organism in which the cells have membrane-bound nuclei, mitochondria, and, in photosynthetic cells, plastids.

eutrophication The process of nutrient enrichment in aquatic environments which occurs naturally over geologic time or is accelerated by activities of the human race.

extracellular matrix Substances, often carbohydrates, produced inside the cell and secreted onto the cell surface external to the plasma membrane.

eyespot A red spot, consisting of carotenoid pigments dissolved in lipids, is responsible for light perception in unicellular and colonial flagellate algae.

filament An algal body, often threadlike, consisting of a linear row of cells in which the neighboring cells share a common cell wall.

fixation A process by which chemical substances are changed from an available form to a less available form. Photosynthesis is a form of carbon fixation.

flagellum (pl. flagella) A complex whiplike structure attached to a microskeleton inside the cell and emergent from the cell that beats to propel the cell forward.

flagellates Cells or colonies that swim propelled forward by their beating flagella.

frond Main part of the algal thallus above the holdfast and stipe, often leaflike in shape.

frustule The boxlike silica cell wall of a diatom.

gametophyte The multicellular gamete-producing phase in algal and plant life histories.

glaucophyte (phylum Glaucophyta) The blue-gray algae whose primitive plastids contain the blue-green and red phycobilin pigments and are bounded by two membranes.

golgi body Cytoplasmic organelle of eukaryotic cells with a secretory function, consisting of a stack of flat disklike tubular membranes which bud off vesicles containing various substances.

haptera The much branched, relatively tough holdfast of kelps.

haptophytes (phylum Haptophyta) Unicellular algae with a haptonema, a threadlike appendage between the two flagella and plastids derived from a red alga.

heterocyst A thick-walled, slightly pigmented cell in some cyanobacteria that is the site of nitrogen fixation.

heterokonts Unicellular algal cells with two dissimilar flagella, a hairy anterior flagellum, and a smooth posterior flagellum.

heterotroph An organism that obtains its nutrition by the uptake of nutrients produced by other organisms.

intercalary meristem A region of cell division in the middle of the thallus.

isogamy (adjective isogamous) A type of sexual reproduction involving morphologically indistinguishable gametes.

macroalgae Algae with thalli that are visible to the unaided eye.

meiosis Cell division that halves the chromosome number from the diploid number (2n) to the haploid number (n) and typically produces four haploid cells.

meristem A cell or a group of cells capable of repeated cell divisions.

microns (micrometer) One micron equals one thousandth (= 0.001) of a millimeter.

mitochondrion A cytoplasmic organelle in eukaryotic cells bounded by a double membrane and responsible for cellular respiration.

mitosis Cell division in eukaryotic cells in which the two progeny are genetically identical to the parent.

mixotroph A mode of nutrition combining the photosynthesis of phototrophs and the consumption of organic matter produced by other organisms (heterotrophy).

monophyletic Used to describe a group of organisms that have descended from a single common ancestor. (*See also* polyphyletic).

mucilage Any of a variety of viscous complex carbohydrates produced by algal cells that are slimy and jellylike when wet.

multiaxial A type of thallus construction composed of many centrally-located (axial) filaments arranged in parallel.

multicellular A thallus composed of many closely adherent cells that have the capacity to communicate with each other and to specialize.

nitrogen fixation The process by which bacteria fix nitrogen gas into ammonia.

oligotrophic Aquatic ecosystems that are low in the nutrients nitrogen, phosphorus, or iron.

oogamy (adj. oogamous) A type of sexual reproduction with a small male gamete (spermatozoid) and a relatively larger nonmotile female gamete (ovum or egg).

parenchyma A relatively undifferentiated tissue consisting largely of cells of more or less equal diameter.

pennate diatom A diatom with bilateral symmetry.

phagocytosis Ingestion of a food item by a cell through the infolding of its plasma membrane that surrounds and encloses the item in a membrane-bound food vacuole inside the cell.

pheromone A chemical compound produced by one organism that acts as a chemical messenger in communication with another organism.

phototroph An organism that obtains its nutrition by photosynthesis.

phototaxis Movement in response to light either toward (positive phototaxis) or away (negative phototaxis) from the light.

pinnate Featherlike branching pattern with two rows of lateral branches either side of the main axis.

plasma membrane The membrane bounding the cell.

plastid The photosynthetic organelle of eukaryotic cells, bounded by plastid membranes and containing thylakoid membranes.

pneumatocyst A gas-filled more or less spherical structure that buoys a thallus toward the water surface.

GLOSSARY

polyphyletic Different species within a group of species that have descended from two or more different ancestors.

primary endosymbiosis The ingestion by a colorless cell of a cyanobacterial cell, which became the plastid of the first algal cell.

primary producer An organism in a food web that obtains its nutrition by photosynthesis.

primary production The rate of production of organic matter (glucose) by photosynthetic organisms (or primary producers).

prokaryote Cells lacking a nucleus and cytoplasmic organelles.

propagule A specialized multicellular structure involved in asexual reproduction.

pyrenoid The protein body in many algae, often associated with the storage of carbohydrate.

protist A single-celled eukaryotic organism.

reactive oxygen species Many different forms (for example superoxide, hydrogen peroxide) formed from molecular oxygen during metabolism that may cause damage to cells.

rhizoids Rootlike unicellular or multicellular filaments responsible for attaching an alga to the substratum, usually a rock surface.

rhodolith Spherical to irregularly shaped nodules of encrusting coralline algae that live unattached on the seafloor.

rhodophytes (phylum Rhodophyta) Red algae that lack flagella, have plastids bounded by two membranes, and contain blue-green and red phycobilin pigments in their plastids.

secondary endosymbiosis The second round of symbiosis, which occurred when a colorless cell ingested a green or red algal cell whose plastid originated from a cyanobacterial cell.

septum (pl. septa) A partition that separates two compartments in a thallus.

siphonocladous A thallus composed of large multinucleate cells.

siphonous A thallus composed of multinucleate tubes or siphons.

sorus (pl. sori) A cluster of reproductive organs, occurring as a patch on the thallus surface.

sulcus Longitudinal furrow of a dinoflagellate cell.

symbiosis A close association between two organisms in which both partners derive some benefits.

thallus (pl. thalli) The relatively undifferentiated algal (and fungal) body that lacks true stems, leaves, and roots.

theca The cell covering of dinoflagellates.

ultrastructure The fine structure of cells as visualized in the electron microscope.

uniaxial A thallus composed of a single central filament.

unicellular An organism composed of a single cell.

vacuole A large fluid-filled sac in a cell, bounded by a membrane.

vegetative cells The nonreproductive cells in the thallus.

vesicle A small often spherical membrane-bound organelle in the cytoplasm of cells.

zooxanthellae Dinoflagellates that live inside the body of coral polyps.

FURTHER READING

Bothwell, J. (2023) *Seaweeds of the World: A Guide to Every Order.* Princeton, NJ: Princeton University Press.

Braume, W., and M.D. Guiry (2011) *Seaweeds: A Colour Guide to the Common Benthic Green, Brown, and Red Algae of the World's Oceans.* Ruggell, Germany: A.R.G. Gantner Verlag KG.

Canter-Lund, H., and J.W.G. Lund (1995) *Freshwater Algae: Their Microscopic World Explored.* Bristol, England: Biopress.

Cavalier-Smith, T. (1998) "A revised six-kingdom system of life." *Biological Reviews*, 73, pp. 203–266.

Graham, L.E., J.M. Graham, and L.W. Wilcox (2009) *Algae* (second edition). San Francisco, CA: Benjamin Cummings.

Krienitz, L. (2018) *Lesser Flamingos: Descendants of the Phoenix.* Berlin, Germany: Springer. (Part One, "The Actors," discusses in detail the flamingo's algal diet.)

Littler, D.S., and M.M. Littler (2000) *Caribbean Reef Plants: An Identification Guide to the Reef Plants of Caribbean, Bahamas, Florida and Gulf of Mexico.* Washington, DC: Offshore Graphics.

Littler, D.S., and M.M Littler (2003) *South Pacific Reef Plants.* Washington, DC: Offshore Graphics.

Thomas, D. (2002) *Seaweeds.* London, England: Natural History Museum.

INDEX

PICTURE CREDITS

The publisher would like to thank the following for permission to reproduce copyright material:

Alamy Stock Photo

2 Nature Picture Library; 3 Steve Speller; 4TL Jennifer Booher; 8 Maidun Collection; 45 Minden Pictures; 66 Artokoloro; 84 Lee Rentz; 86 Leo Francini; 91 Mark Conlin; 93 Jennifer Booher; 121 Andi Edwards; 151 agefotostock; 155 Premaphotos; 161 Minden Pictures; 166–167 jonathan nguyen; 172T Stephan Schramm; 172B Barbara von Hoffmann; 174 FLPA; 176–177 Cattie Coyle; 194 Blue Planet Archive; 199 Alessandro Mancini; 211 Bob Gibbons; 215 imageBROKER; 217 Design Pics Inc; 219 Nature Picture Library; 227 Auscape International Pty Ltd; 234 imageBROKER; 235 Steve. Trewhella; 238 Javier Etcheverry; 239B Jennifer Booher; 240 Joerg Boethling; 243 Artokoloro; 245 Genevieve Vallee; 246 Joaquin Ossorio-Castillo; 247 Nir Alon; 248 BIOSPHOTO; 250L cbimages; 252 Nature Picture Library; 258 Daniel Poloha Underwater; 267 Steve Speller; 273 Alison Thompson;

Getty Images

180 Auscape

Minden Pictures

113 Mathieu Foulquie

Nature Picture Library

82 Nick Upton; 88 Georgette Douwma; 149 Ingo Arndt; 157 Nick Hawkins; 168–169 Alex Hyde; 190 Sue Daly; 191 Richard Robinson; 196 Nick Upton; 213 Shane Gross; 249 Alex Mustard; 259 Alex Mustard;

Non-agency

5TR © M.D. Guiry, AlgaeBase; 5BL Julie A. Phillips; 5BC David Domozych, Department of Biology, Skidmore College; 5BR Professor Mitsunobu KAMIYA; 7 Daniel Solander Library, Botanic Gardens of Sydney; 9 Roger Steene; 10–11 Dr. Jeremy R. Young, University College London; 12–13 Neville Coleman; 16–17 B Christopher; 21R Neville Coleman; 24 reproduced with permission of the Freshwater Biological Association on behalf of The Estate of Dr Hilda Canter-Lund; 25 reproduced with permission of the estate of Lynn Margulis; 33 Roger Steene; 35TL, TC & TR N. J. Butterfield, University of Cambridge; 35B Pavel Škaloud; 36 David Domozych, Department of Biology, Skidmore College; 40–41 Neville Coleman; 43 Julie A. Phillips; 47 NOAA; 49, 53 Reproduced with permission of the Freshwater Biological Association on behalf of The Estate of Dr Hilda Canter-Lund; 55 CC BY 3.0/CSIRO Science Image Library; 59 Reproduced with permission of the Freshwater Biological Association on behalf of The Estate of Dr Hilda Canter-Lund; 63 David Williamson; 78 Claude Payri/IRD/MooreaBiocode; 90 Neville Coleman; 98 Roger Steene; 99 Neville Coleman; 105, 107 © M.D. Guiry, AlgaeBase; 109 Photography by Paul W. Gabrielson; 111 Dr. John Huisman; 129 Dr. John Huisman; 136B Dr. John Huisman; 139 David G.

Mann, Royal Botanic Garden Edinburgh; 140 Ms Alisa Mihaila; 141 Dr. John Huisman; 147 Professor Dr. Peter Wirtz; 153 Julie A. Phillips; 159 © M.D. Guiry, AlgaeBase; 165 Michiel Vos, Instagram: @an_bollenessor; 168L CC BY 2.0/Dick Culbert; 170 Dr. Alf Skovgaard; 181 Professor Mitsunobu KAMIYA; 183 NOAA/photo courtesy of Mike Echevarria, Florida Aquarium; 184 Ignacio Manuel Bárbara Criado; 185 Roger Steene; 186T CC BY 4.0/NOAA/NMFS/PIFSC/CRED, Oceanography Team; 186B Dr. Maggie Johnson; 187 Neville Coleman; 188–189 Roger Steene; 193 Patrick Martone, www.botany.ubc.ca/martone, Twitter: @martonelab; 198 Image courtesy of FGBNMS/UNCW-NURC/NOAA; 201 Hans J. Sluiman; 203 Albert Calbet; 204 Photographer Dave Reynolds "As above, so below," South Arm, Tasmania; 206 CC BY-SA 2.5/NEON; 208 Robin J. Fales; 221 Image courtesy of FGBNMS/UNCW-NURC/NOAA; 223 Australian Institute of Marine Science/photographer Dr. James Gilmour; 225 Professor Myung Park; 230 NASA Earth Observatory images by Joshua Stevens, using Landsat data from the U.S. Geological Survey, and MODIS data from NASA EOSDIS/LANCE and GIBS/Worldview; 232 NASA image by Norman Kuring, NASA's Ocean Color Web; 242 CC BY 4.0/Marine Biological Laboratory (Woods Hole, Mass.) Digitized by MBLWHOI Library; 251 Professor Mitsunobu KAMIYA; 255 Miriam Godfrey/NIWA; 261 Dr. Jeremy R. Young; 275 Scottish Association for Marine Science (SAMS); 277 CC BY-NC-ND 4.0/Miguel de Salas

Science Photo Library

4TR Rogelio Moreno; 4B Dr Keith Wheeler; 4CR Wim Van Egmond; 18 Gerd Guenther; 51 Wim Van Egmond; 57 Martyn F. Chillmaid; 68 Gerd Guenther; 71 Biophoto Associates; 72 Gerd Guenther; 73 Wim Van Egmond; 74 Gerd Guenther; 75, 77 Wim Van Egmond; 79 Dr Keith Wheeler; 94 Steve Gschmeissner; 101 Natural History Museum, London; 117 Ted Kinsman; 122 Sinclair Stammers; 123 Pedro Neves; 125 Copyright holder Julie Phillips, photographer Kay Abel; 126–127 M.I. Walker; 133 Dr Keith Wheeler; 136T Carolina Biological Supply Company; 145 Gerd Guenther; 178 Rogelio Moreno; 204 (inset) Wim Van Egmond; 207 Patrice Latron/Look At Sciences; 236 Gerd Guenther; 244 Pascal Goetgheluck; 263 Biophoto Associates; 269 Frank Fox; 271 Sinclair Stammers

Shutterstock

20–21 Benny Marty; 38–39 Four candles; 81 Ashish_wassup6730; 83 aquapix; 87 KPG-Payless; 103 Lebendkulturen.de; 115 Philip Garner; 171 Lam Van Linh; 179 Sirisak_baokaew; 197 daguimagery; 237 Marko5; 239T Kongsak; 241 Ventura; 250RT triocean; 250RB Fahroni; 253 Olga Kozyr; 256 Jillian Cain Photography; 265 Gold33

All reasonable efforts have been made to trace copyright holders and to obtain their permission for the use of copyright material. The publisher apologizes for any errors or omissions in the list above and will gratefully incorporate any corrections in future reprints if notified.

ACKNOWLEDGMENTS

This book was inspired by the last six decades of ground-breaking scientific research that has revolutionized our understanding of algal evolution and biology. Two generations of phycologists worldwide—far too many to acknowledge here—have mounted an extraordinary research effort, which has culminated in an "algal revolution." Their sensational discoveries have been published in the scientific literature, which, for the general reader, is often inaccessible and difficult, if not impossible, to read. In *The Lives of Seaweeds*, I have endeavoured to communicate the exciting science of the algal world to the general reader as well as providing them with the knowledge that algae matter—their existence and ours are inextricably linked.

I greatly appreciate, and am overwhelmed by, the generosity of numerous enthusiastic colleagues who have provided their extraordinary images for this book, most often with the simple desire to help me share the delight of this sometimes hidden world with a broad popular audience. The images they have supplied have greatly enhanced the scientific and aesthetic content of the book. I sincerely thank Kay Abel, Dr. Albert Calbet, Robin Fales, Dr. Paul Gabrielson, Dr. James Gilmour, Professor Gustaaf Hallegraeff, Professor David John, Professor Mitsunobu Kamiya, Professor David Mann, Ailsa Mihaila, Pedro Neves, Professor Myung Gil Park, Dave Reynolds, Dr. Alf Skovgaard, Dr. Hans Sluiman, Michiel Vos, David Williamson, Professor Dr. Peter Wirtz, and Dr. Jeremy Young for providing their outstanding images. Many of these photographs were captured by academic researchers and have never previously been seen outside scientific circles. I am awe-struck by their beautiful photographs, and I marvel at the technical difficulties they had to overcome. Each picture represents days, months, or years of patient work to capture some unique and extraordinary images.

I must also give special mention and gratitude to the talented Australian underwater photographers Roger Steene and the late Neville Coleman, who generously gave permission to freely use what I needed from their extensive libraries of seaweed images.

Some colleagues went above and beyond to provide images. Miguel Garcia, the librarian at the Royal Botanic Garden, Sydney, and the garden's honorary phycologist, Dr. Steve Skinner, sent me title page images from two Victorian-era seaweed albums. The image, which I didn't know existed, made a wonderful addition to this book. Marie Roman of the Australian Institute of Marine Science's Visual Content Production liaised with coral researcher Dr. James Gilmour to provide a great coral bleaching image, for which James then also kindly provided helpful scientific interpretation. Dr. John Huisman of the Western Australian Herbarium graciously sorted through his image collection on many occasions looking for the images I needed.

One scientist, my esteemed colleague and dear friend Peter Davie, has made a huge contribution to this book and, in doing so, has earned my eternal gratitude. Over the decades, Peter and I have worked together on field guides for marine life and on marine projects in Australia, including as co-convenors of an international marine research workshop. Despite not being a phycologist, but a renowned crustacean taxonomist, the wonderfully supportive Peter reached out to his photographer friends to help me find rare algal images that could never be obtained in commercial image databases. In particular, his close friend Roger Steene generously gave us his large collection of seaweed images. Peter also spent many hours using his considerable graphics experience to edit and upgrade many images to the high standard required for publication. Numerous stunning images are only in the book because of Peter's huge contribution to this project.

The Lives of Seaweeds: A Natural History of our Planet's Seaweeds and other Algae is an ambitious book, and possibly the only book for a general audience that deals with organisms from the entire algal world—four of the six kingdoms of life! Thus, the commissioning editor, Kate Shanahan, must be thanked for her enlightened vision that such a book could be achieved. The editorial team at Unipress, Kate Duffy, Lesley Henderson, and Kathleen Steeden, must also be thanked for their editing skills that have helped bring the book through to fruition. Finally, thank you to Princeton University Press for publishing this much-needed introduction to the entire algal world.